AUTOCAD 2014
BEGINNING
AND
INTERMEDIATE

AutoCAD 2014
Beginning
and
Intermediate

By
Munir M. Hamad
Autodesk™ Approved Instructor

Mercury Learning and Information
Dulles, Virginia
Boston, Massachusetts
New Delhi

Publisher: David Pallai

MERCURY LEARNING AND INFORMATION
22841 Quicksilver Drive
Dulles, VA 20166
info@merclearning.com
www.merclearning.com
1-800-758-3756

This book is printed on acid-free paper.

Munir M. Hamad, AUTOCAD™ 2014 BEGINNING AND INTERMEDIATE.
ISBN: 978-1-938549-62-5

The publisher recognizes and respects all marks used by companies, manufacturers, and developers as a means to distinguish their products. All brand names and product names mentioned in this book are trademarks or service marks of their respective companies. Any omission or misuse (of any kind) of service marks or trademarks, etc. is not an attempt to infringe on the property of others.

Library of Congress Control Number: 2013944479
131415321

Our titles are available for adoption, license, or bulk purchase by institutions, corporations, etc. For additional information, please contact the Customer Service Dept. at 1-800-758-3756 (toll free).

CONTENTS

ABOUT THE BOOK

This book is the most comprehensive book you will find on AutoCAD 2014 – 2D Drafting. It is divided into three major parts. The first, Essentials, covers ten chapters and three projects (one architectural and two mechanical using both imperial and metric units). The second part, Intermediate, contains eight chapters. It contains topics covered in the first section, is more depth, or additional knowledge needed to fill in the gaps left in the first part. The final part of the book, Advanced, contains seven chapters discussing the most advanced features of AutoCAD 2014.

If you don't have any prior experience using AutoCAD, you can start with any chapter of the book. But if you want to be an advanced AutoCAD user, you should go through all 25 chapters and complete all the projects and practices.

This book can also help you prepare for the *AutoCAD Certified Professional* exam, given by Autodesk, Inc.

This book's chapters are divided as follows:

- Chapter 1 covers the basics of AutoCAD along with the interface.
- Chapter 2 covers AutoCAD techniques for drawing with accuracy.
- Chapters 3 & 4 cover the modifying commands used to modify and construct drawings.
- Chapter 5 covers layers and inquiry commands.
- Chapter 6 covers creating and editing blocks and inserting and editing hatches.
- Chapter 7 covers AutoCAD methods for writing text.
- Chapter 8 covers how to create and edit dimensions in AutoCAD.

- Chapter 9 covers how to plot your drawing.
- Chapter 10 includes three projects, one architect and two mechanical covering both metric and imperial units.
- Chapter 11 covers more 2D objects creation.
- Chapters 12 & 13 cover advanced practices and techniques.
- Chapter 14 covers block tools and block editing.
- Chapter 15 covers text styles and table styles along with formulas in tables.
- Chapter 16 covers dimension styles and multileaders.
- Chapter 17 covers plot styles, the meaning of Annotative, and creating DWF files.
- Chapter 18 covers how to create a template file and customizing the AutoCAD interface.
- Chapter 19 covers parametric constraints.
- Chapter 20 covers dynamic blocks.
- Chapter 21 covers block attributes.
- Chapter 22 covers Xref.
- Chapter 23 covers sheets sets.
- Chapter 24 covers CAD standards and advanced layer commands.
- Chapter 25 covers the drawing review.

PREFACE

- Since its inception, AutoCAD has enjoyed a very wide user base and has been the most widely used CAD software since the 1980s. This popularity is due to its logic and simplicity, which makes it very easy to learn.
- This book addresses all levels of AutoCAD 2D drafting: Essentials, Intermediate, and Advanced AutoCAD techniques. It is not a replacement for the manual(s) that comes with the software, but is considered complementary with its practices and projects, meant to strengthen the knowledge gained and solidify the techniques discussed.
- Solving all practices is essential because AutoCAD is a practical tool and not theoretical.
- At the end of each chapter you will find "Chapter Review Questions." These are the same sort of questions you might see in an Autodesk exam. The answers to the odd questions are included at the end of each chapter to check your work.
- Chapter 10 contains three projects: One for an architectural plan and the other two for mechanical engineering. (One of these is explained in detail, so you can follow the steps by themselves without any help.) Solving these will allow you to master the knowledge needed to land a job in today's market. These projects are presented in metric and imperial units.

ABOUT THE DVD

- The DVD included with this book contains:
  - A link to the AutoCAD 2014 Trial version, which will last for 30 days starting from the day of installation. This version will help you solve all the exercises and workshops in this book. *The student trial version can be extended beyond the 30 day period.*
  - The practices files which will be your starting point to solve all exercises and workshops in the book.
  - Copy the folder named "Practices and Projects" onto the hard drive of your computer. In the Project folder, you will find two folders; the first is called "Metric" for metric units projects and the second one is called "Imperial" for imperial units projects.

Prerequisites

- The author assumes the reader has experience using computers and the Windows operating system. You should know basic commands and be able to create new files, open existing ones, save and use save as well as close and exit the software. These fundamentals are not covered in this book.
- AutoCAD 2014 comes with a dark gray background, but the screenshots in this book have been changed to white to create clearer illustrations.

Chapter 1

AutoCAD 2014 Basics

In This Chapter

◇ How to start AutoCAD
◇ How to work with the AutoCAD interface
◇ AutoCAD defaults and drawing units
◇ How to work with file oriented-commands
◇ Undo and Redo commands

1.1 HOW TO START AUTOCAD

- AutoCAD was released in 1982 by Autodesk, Inc. (a small company at the time) and designed for PCs only. Since then AutoCAD has boasted the biggest user base in the world in the CAD industry. You can use AutoCAD for both 2D and 3D drafting and designing and for architectural, structural, mechanical, electrical, road and highway designs, environmental, and manufacturing drawings.
- Though the theme these days is BIM (Building Information Modeling) AutoCAD is still the most profitable software for Autodesk, Inc. due to its ease-of-use and totality.
- There is another version of AutoCAD, called AutoCAD LT, used only for 2D drafting.
- To start AutoCAD 2014, double-click the shortcut on your Desktop (created in the installation process), and AutoCAD will show the **Welcome** window which looks like the following:

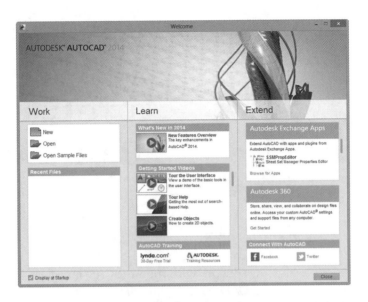

- Using the Welcome window, you can create a new file, open an existing file, open a sample files, or open recent files, and you can even watch tutorial videos. The Welcome window also offers a 30-day free trial with Lynda.com, and allows you to view some Autodesk Training Resources. You can also download apps for AutoCAD, start a new account with Autodesk 360 (Autodesk Cloud Services), and connect to AutoCAD via Facebook and Twitter.
- The AutoCAD 2014 interface contains the following parts:

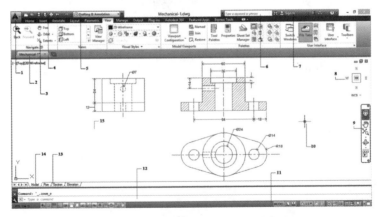

1. Application menu	2. File tab	3. Ribbon
4. Quick access toolbar	5. Workspace	6. Info center
7. Autodesk 360	8. ViewCube	9. Navigation bar
10. Cross hairs	11. Status bar	12. Command window
13. Layout tab	14. Model tab	15. Graphical area

1.2 AUTOCAD 2014 INTERFACE

This interface is similar to the Microsoft Office 2007/2010 interface. You will mainly use the Ribbons and Application Menu to reach commands. The most important feature of this interface is the size of the **Graphical Area**, which is much larger.

1.2.1 Application Menu

- The Application Menu contains the file-related commands:

- Commands such as creating a new file, opening an existing file, saving the current file, saving the current file under a new name and different folder, exporting the current file to a different file format, printing and publishing the current file, etc., are found here. Most of these commands will be discussed throughout the book. By default, you will see your recent files, and you can choose how to display them in the Application Menu using this control:

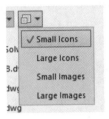

- You can also decide how to sort your recent files using this control:

1.2.2 Quick Access Toolbar

- This toolbar contains all the File commands along with Workspace and Undo/Redo:

- You can customize this toolbar by clicking the arrow; you will see the following:

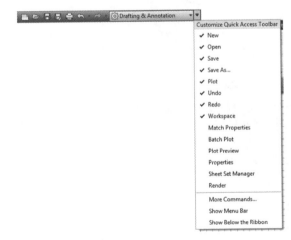

- As you can see you can add or remove commands and choose **Show Menu Bar**, which you may sometimes find useful, since the Ribbons do not include all the AutoCAD commands.

1.2.3 Ribbons

- Ribbons consist of two parts: tabs and panels, as shown below:

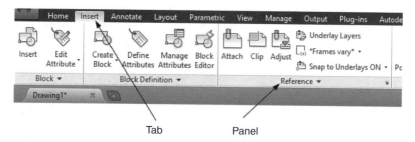

         Tab   Panel

- Some panels have more buttons than shown. The following is the **Modify** panel:

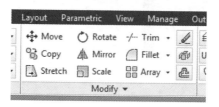

- Click the small triangle near the title, and you will see the following:

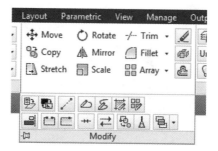

- If you move away from the panel, the buttons will disappear. To make them visible again click the push pin, and the new view will look like the following:

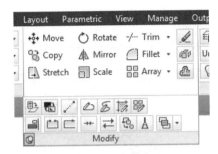

- For some commands, there are many options. To make things easier, AutoCAD put all the options with the corresponding button. See the following illustration:

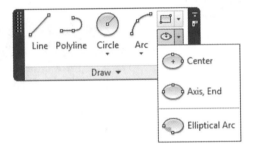

- Ribbons have a very simple but useful feature. If you move the mouse over a button a small help screen will pop up:

- If you leave the mouse pointer over the button, AutoCAD will show more detailed help:

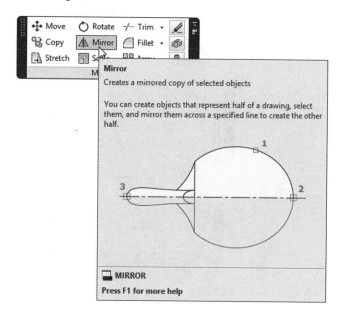

- Other buttons have video help, like the following:

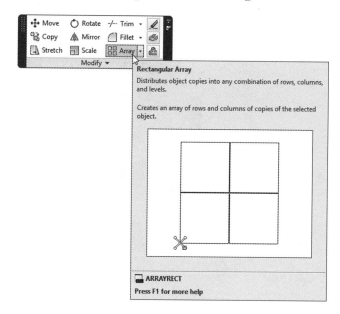

- Panels have two states, *docked* or *floating*. By default, all panels are docked in their respective tab. Drag and drop the panel in the graphical area to make it floating. If a panel is floating, you will be able to see it while other tabs are active.
- You can send the panel back to its respective tab by clicking the small button at the top-right side:

- While the panel is floating you can toggle the orientation:

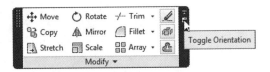

- It will either extend to the right:

- Or down:

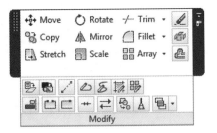

- The small arrows at the end of the tabs allow you to cycle through the different states of the Ribbons. The main objective of this new feature is to give you yet more graphical area to work in. Clicking the small arrow will show the following:

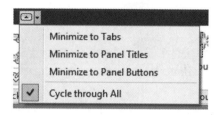

1.2.4 InfoCenter

- The **InfoCenter** is the place to find help topics online and offline, along with other helpful tools:

- For example, if you type a word or phrase in the field shown, AutoCAD will open the Autodesk Exchange window and find all the related topics online and offline. Online means it will search all Autodesk websites along with some popular blogs.
- Sign In will allow you to sign into Autodesk Online Services. The X at the right will activate the Autodesk Exchange window and show you the latest information and video tutorials. The last button at the right with the question mark will show the following:

1.2.5 Command Window

▪ Reading the Command window will help you understand what Auto-CAD wants from you. AutoCAD will show two things in the Command window: your commands and AutoCAD prompts asking you to do something such as specify a point, an angle, etc. See the following illustration:

```
X  Arc creation direction: Counter-clockwise (hold Ctrl to switch direction).
      ARC Specify start point of arc or [Center]:
```

1.2.6 Graphical Area

▪ The graphical area is your drafting area. This is where you will draw all your lines, arcs, circles, etc. It is a precise environment with X,Y,Z space for 3D and the X,Y plane for 2D. You can monitor coordinates in the left part of the status bar.

1.2.7 Status Bar

▪ The status bar in AutoCAD contains coordinates along with important functions; some of the functions are for precise drafting and some of them are for other more advanced features such as Infer Parameters, Cycling, and Transparency.

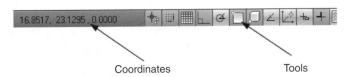

Coordinates   Tools

1.3 AUTOCAD DEFAULTS

▪ There are some settings you should be familiar with before working in the AutoCAD environment, so keep the following in mind:
  • AutoCAD saves points as Cartesian coordinates (X,Y) for both metric and imperial numbers. This is the first method of precise input in AutoCAD, typing the coordinates using the keyboard:

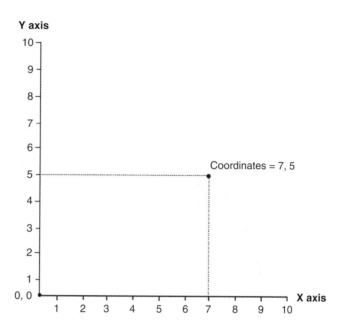

- To specify angles in AutoCAD, assume East (to your right) is your 0° and then go Counter-Clockwise. (This is applicable only for the northern hemisphere, but we will learn in Chapter 3 how we can change this setting for the southern hemisphere.)

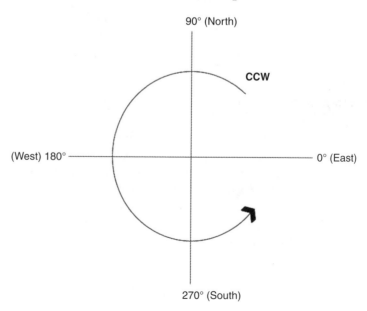

- The mouse wheel has four zooming functions: Zoom In (move wheel forward), Zoom out (move wheel backward), Panning (press and hold the wheel), and Zooming Extents (double-click the wheel).
- Pressing [Enter] or [Spacebar] is the same in AutoCAD.
- Pressing [Enter] without typing any command in AutoCAD will repeat the last command. If it is the first thing you do in the current session, it will start Help.
- Pressing [Esc] will cancel the current command.
- Pressing [F2] will show the Text Window. See the following:

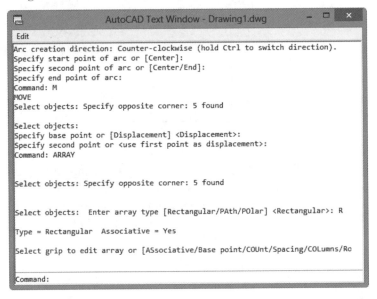

1.4 DRAWING UNITS

- If you draw a 6-unit line in AutoCAD, what units will AutoCAD use? Will it be 6 m or 6 ft, or neither? AutoCAD works with all types of units; if you want to use 6 m, AutoCAD will do this. If you mean 6 ft AutoCAD will work with this unit as well. You just need to be consistent with your measurements throughout the file. While this is true in Model Space where you will do your drafting, when it comes to printing, you will have to make sure to set your drawing scale correctly. In Chapter 9, we will discuss printing.

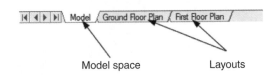

Model space   Layouts

1.5 CREATING A NEW AUTOCAD DRAWING

- This command allows you to create a new drawing based on a pre-made template (we will discuss how to create your own template in Appendix A). Use the **Quick Access Toolbar** and click the **New** button:

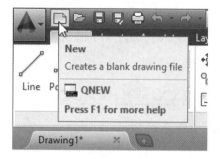

- Or you can click the (+) sign in the File tab:

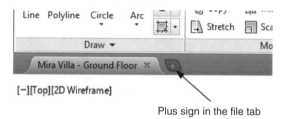

Plus sign in the file tab

- You will see the following dialog box:

- Take the following steps:
  - Select the desired template file (AutoCAD template files have an **.dwt* extension). AutoCAD 2013 comes with many pre-made templates you can use (it is actually better, however, to create your own template files).
  - Once you are done, click the **Open** button.
  - AutoCAD drawing files have a **.dwg* file extension.
  - AutoCAD will start with a new file with a temporary name such as *Drawing1.dwg*, and you should rename it to something meaningful.

1.6 OPENING AN EXISTING AUTOCAD DRAWING

- This command allows you to open an existing drawing file for additional modifications. From the **Quick Access** Toolbar, click the **Open** button:

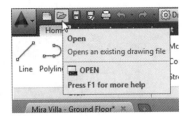

- You will see the following dialog box:

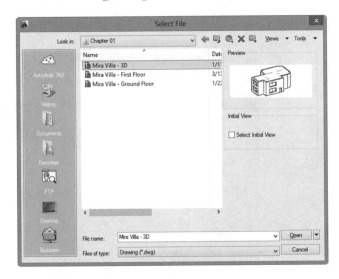

- Take the following steps:
  - Specify your desired drive and folder.
  - You can open a single file by selecting its name from the list and clicking the **Open** button, or you can double-click on the file's name. Or, you can open more than one file by selecting the first file name, then holding the [Ctrl] key and clicking the other file names in the list (a common Windows shortcut) and then clicking the **Open** button.

1.6.1 File Tab

- Using the File tab beneath the ribbon, you will see for each opened file a tab, just like the following:

- The color used is gray. The current file (tab) will be a lighter gray, but the other tabs will be a darker gray. If you hover over one of the tabs, two things will take place:
  - The path of the file will appear above the tab
  - The model space and the layouts of the file will appear beneath
- See the following:

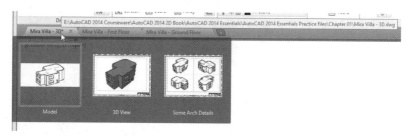

- There will also be a blue frame around the Model Space view. Moving your mouse to the right will show at the graphical area the layouts; if you find what you are looking for, click the layout view to move to it.

NOTE

  - A star beside the name of the file in the file tab means this file was changed and you need to save the changes.
  - Click (x) beside the name to close the file, hence closing the tab.

1.7 CLOSING DRAWING FILE(S)

- This command allows you to close the current opened file(s), or all opened files depending on the command you choose. Use the **Application Menu** and move your mouse to the **Close** button, then select either **Current Drawing** to close the current file or **All Drawings** to close all the opened files with a single command:

- If any of the opened files was modified, AutoCAD will ask if you want to save or close without saving. See the following dialog box:

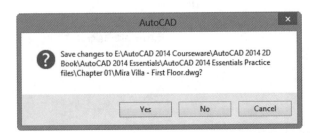

1.8 UNDO AND REDO COMMANDS

- Undo and Redo help you to correct mistakes. They can be used in the current session only.

1.8.1 Undo Command

- This command will undo the effects of the last command. You can reach this command by going to the **Quick Access toolbar** and clicking the **Undo** button. If you want to undo several commands, click the small arrow at the right. You will see a list of the commands; select the group and undo them:

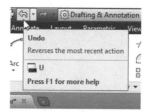

- You can also type **u** at the Command window (don't type **undo** because it has a different meaning here), or press [Ctrl] + Z on the keyboard.

1.8.2 Redo Command

- This command will undo the Undo command. You can reach this command from the **Quick Access toolbar** and clicking the **Redo** button. If you want to redo several commands, click the small arrow at the right. You will see a list of the commands; select the group and redo them:

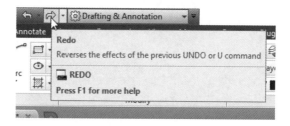

- You can also type **redo** at the Command window, or press [Ctrl] + Y on the keyboard.

PRACTICE 1-1

 AutoCAD Basics

1. Start AutoCAD 2014.
2. Open the following files:
   a. Mira Villa – Ground Floor
   b. Mira Villa – First Floor
   c. Mira Villa – 3D
3. Using File tab, take a look at the three files and their layouts.
4. Use the different zoom techniques with the mouse wheel.
5. Using the Application Menu, close all files without saving.

NOTES:

CHAPTER REVIEW

1. File tab allows you to see the Model Space and layouts of open files.
   a. True
   b. False
2. AutoCAD template files have the _____ extension.
   a. *.dwt
   b. *.dwg
   c. *.tmp
   d. *.temp
3. AutoCAD units can be meters, or feet, which you choose.
   a. True
   b. False
4. Moving the mouse wheel forward will _____.
5. To undo any command in AutoCAD you can:
   a. Click the Undo icon from Quick Access Toolbar
   b. Type **u** at the Command window
   c. Use [Ctrl] + Z
   d. All of the above
6. Ribbons consist of _____ and _____
7. The menu bar is not shown by default, but you can make it visible.
   a. True
   b. False
8. The AutoCAD drawing file extension is _____.
9. Positive angles in AutoCAD are _____.

CHAPTER REVIEW ANSWERS

1. a
3. a
5. d
7. a
9. CCW

Chapter **2**

PRECISE DRAFTING IN AUTOCAD 2014

In This Chapter

◇ Drafting priorities
◇ How to draw lines, circles, and arcs using precise methods
◇ How to draw polylines using precise methods
◇ How to convert lines and arcs to polylines and vice-versa
◇ OSNAP and OTRACK

2.1 DRAFTING PRIORITIES

- There are two main drafting priorities for most people: accuracy and speed. Most people want to finish their drawings fast, but without compromising on accuracy. Experts tend to put accuracy first, but at the expense of speed.
- In this chapter, you will learn how to use the four most important drafting commands. These are:
  - The Line command, used to draw line segments.
  - The Arc command, used to draw circular arcs.
  - The Circle command, used to draw circles.
  - The Polyline command, used to draw lines and arcs jointly.
- While we are discussing the four drafting commands, we will also cover accuracy tools and tools to help you speed up the drafting process.

2.2 DRAWING LINES USING THE LINE COMMAND

- This command allows you to draw straight lines; each line segment pre-
  sents a single object. To issue this command, go to the **Home** tab, locate
  the **Draw** panel, then select the **Line** button:

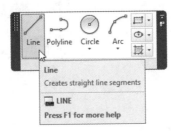

- You will see the following AutoCAD prompts:

```
Specify first point:
Specify next point or [Undo]:
Specify next point or [Undo]:
Specify next point or [Close/Undo]:
```

- Using the first prompt specify the coordinates of your first point. Keep
  specifying points until you are done, and keep the following in mind:
  - If you want to stop without closing the shape, simply press [Enter].
    ([Esc] will do the job as well, but don't make it a habit, since [Esc]
    generally means abort.)
  - If you want to close the shape and finish the command, press **C** on the
    keyboard, or right-click and select the **Close** option.
  - If you make a mistake you can undo the last point by typing **u** on the
    keyboard, or right-clicking and selecting the **Undo** option.
- This is what the right-click menu looks like:

2.3 WHAT IS DYNAMIC INPUT IN AUTOCAD?

- Dynamic Input has a couple of functions:
  - It will show all the prompts at command window in the graphical area
  - It will show lengths and angles of the lines before drafting, which will allow you to specify them accurately
- In order to turn on/off the **Dynamic Input** use the following button in the status bar:

2.3.1 Example for Showing Prompts

- By default, if you type any command using the Command window, AutoCAD will help you by showing all the commands starting with the same word(s). See the following example:

- We typed the two letters **LI** and AutoCAD gave us all the commands starting with these two letters. While Dynamic Input is on, this is applicable to the crosshairs as well. You will see the following:

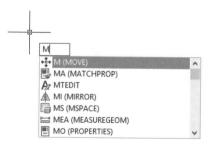

- Select the Line command from the list. Once you press [Enter], the following prompt will appear:

- Type in the X and Y coordinates using the [Tab] key to move between the two fields:

2.3.2 Example for Specifying Lengths and Angles

- Once you have specified the starting point, AutoCAD will use Dynamic Input to show the length and the angle of the line using **Rubberband** mode:

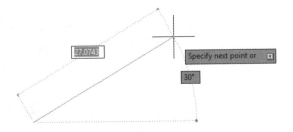

- **NOTE** Angles are measured CCW starting from the East, but only for 180°, unlike the angle system in AutoCAD, which is for the whole 360°.
- Type the length of the line, then using [Tab] input the angle (it will increase in 1° increments). Once you are finished, press [Enter] to specify the first line, and continue doing the same for the other segments:

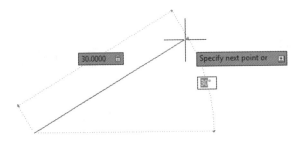

PRACTICE 2-1

 Drawing Lines Using Dynamic Input

1. Start AutoCAD 2014.
2. Open **Practice 2-1.dwg**.
3. Using the status bar, click off Polar Tracking, Ortho, Object Snap, and make sure Dynamic Input is on.
4. Draw the following shape, using 0,0 as your starting point, keeping in mind all sides = 4 and all angles are multiples of 45°:

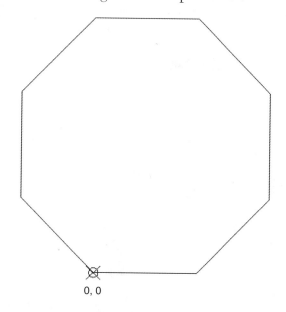

0, 0

5. Save the file and close it.

2.4 EXACT ANGLES (ORTHO VS. POLAR TRACKING)

- Dynamic Input angles are incremented by 1°, but you cannot depend on Dynamic Input to specify angles precisely in AutoCAD.
- The **Ortho** function will force the lines to be right angles (orthogonal) using the following angles: 0, 90, 180, and 270.

- In order to turn on/off **Ortho**, use the following button on the status bar:

- But what if you want to use other angles such as 30, 45, 60, etc.? For this reason AutoCAD introduced another function called Polar Tracking, which allows you to see in the graphical area <u>rays</u> starting from the current point pointing towards angles such as 30, 45, etc., and based on the settings, you can specify angles. Since Ortho and Polar Tracking contradict each other, when you switch to one, the other will be turned off automatically.
- To turn on/off **Polar Tracking**, use the following button on the status bar:

- When you right-click the button on the status bar, you will see the following menu:

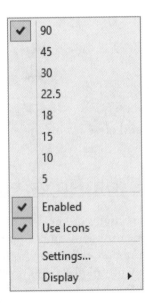

- You can select the desired angle, or select **Settings** to change some of the default settings of Polar Tracking. You will see the following dialog box:

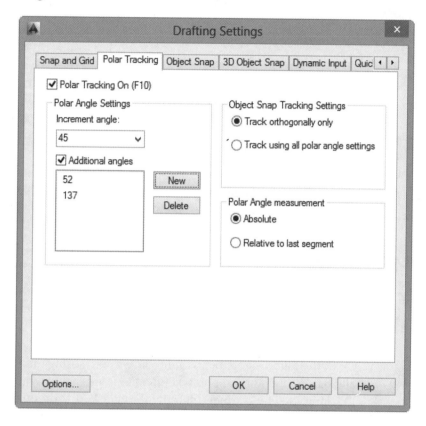

2.4.1 Increment Angle

- The Increment angle is the angle to be used along with its multiples. Select one from the list or type your own.

2.4.2 Additional Angles

- If you are using 30 as your increment angle, then 45 will not be among the angles that Polar Tracking will allow you to use. So, you will need to specify it as an additional angle. But be aware that you will not use its multiples.

2.4.3 Polar Angle Measurement

- When you are using Polar Tracking, you have the ability to specify angles as absolute angles (based on 0° at the East) or use the last line segment as your 0 angle. See the following illustration:

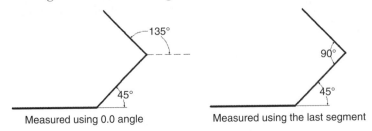

Measured using 0.0 angle   Measured using the last segment

- While you control the angle using either Ortho or Polar Tracking, you can type in the distance desired, then press [Enter]. This allows you to draw accurate distances. This method is called *Direct Distance Entry*.

PRACTICE 2-2

Exact Angles

1. Start AutoCAD 2014.
2. Open **Practice 2-2.dwg**.
3. Draw the following shape (without dimensions) using the Line command and 0,0 as your starting point, keeping in mind that you have to use Polar Tracking. Set the proper Increment angle and additional angles using the Direct Distance Entry method to input the exact distances:

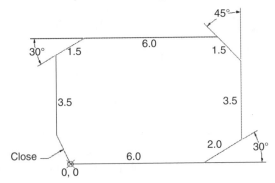

4. Save and close the file.

2.5 PRECISE DRAFTING USING OSNAP

- Object Snap, or OSNAP, is the most important accuracy tool in AutoCAD for 2D and 3D as well. It is a way to specify points on objects precisely using the AutoCAD database stored in the drawing file.
- Some OSNAPs include:
  - **Endpoint**: To catch the Endpoint of an object (line, arc, or polyline).
  - **Midpoint**: To catch the Midpoint of an object (line, arc, or polyline).
  - **Intersection**: To catch the Intersection of two objects (any two objects).
  - **Perpendicular**: To catch the Perpendicular point on an object (any object).
  - **Nearest**: To catch on an object Nearest to your click point (any object).
- We will discuss more OSNAPS when we discuss more drawing objects. Here are some graphical presentations of each one of these OSNAPs:

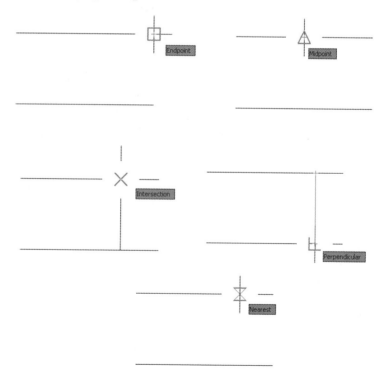

2.5.1 Activating Running OSNAPs

- To activate running OSNAP in the drawing, click the **Object Snap button on the status bar:**

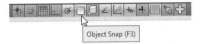

- When you switch this button on, you will use whatever OSNAPs were on. In order to use your own, right-click the **Object Snap** button and the following menu will appear:

- You can switch on the desired OSNAPs one-by-one.
- You can also select the **Settings** option and the following dialog box will appear:

- There are two buttons at the right: **Select All** and **Clear All**. We recommend using Clear All and then selecting the desired OSNAPs. When done click **OK**.

2.5.2 OSNAP Override

- While the **Object Snap** button is on, some OSNAPs are working and others are not. Sometimes you may want to switch them all off and only use one; other times, you may want to switch everything back to normal. This is what we call OSNAP Override.
- There are two ways to activate an override:
  - Using the keyboard, type the first three letters of the desired OSNAP.
  - Using the keyboard, hold the [Shift] key and then right-click. You will see the following pop-up menu:

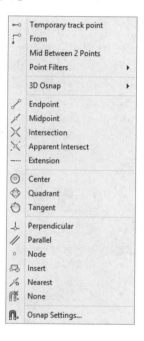

```
 ↦    Temporary track point
 ↦°   From
      Mid Between 2 Points
      Point Filters          ▸
      3D Osnap               ▸
 ⌁    Endpoint
 ⌁    Midpoint
 ✕    Intersection
 ✕    Apparent Intersect
 ----- Extension
 ⊙    Center
 ⬙    Quadrant
 ◌    Tangent
 ⊥    Perpendicular
 //   Parallel
 ∘    Node
 ⊡    Insert
 ⽊    Nearest
 ♒    None
 ♒.   Osnap Settings...
```

PRACTICE 2-3

Object Snap (OSNAP)

1. Start AutoCAD 2014.
2. Open **Practice 2-3.dwg**.
3. Check OSNAP at the status bar and make sure that only Endpoint and Midpoint is switched on.

4. Using OSNAP draw lines in the drawing to make it look like the following:

5. Save and close the file.

2.6 DRAWING CIRCLES USING THE CIRCLE COMMAND

- This command will draw a circle using different methods based on the available data. If you know the coordinates of the center, there are two available methods. If you know the coordinates of points at the parameter of the circle, there are two additional methods. Finally, if there are drawn objects such as lines, arcs, or other circles that can be used as tangents for the to-be-created circles, there are two more methods. There are six methods for drawing a circle in AutoCAD:

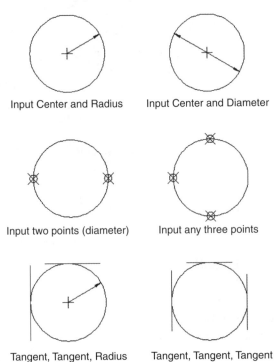

Input Center and Radius   Input Center and Diameter

Input two points (diameter)   Input any three points

Tangent, Tangent, Radius   Tangent, Tangent, Tangent

- To issue this command, go to the **Home** tab, locate the **Draw** panel, then select the arrow near the **Circle** button to see all the available methods:

2.7 DRAWING CIRCULAR ARCS USING THE ARC COMMAND

- This command will draw an arc part of a circle. To make our lives easier, AutoCAD allows you to use eight pieces of information related to a circular arc. These are:
  - The starting point of the arc.
  - Any point as a second point on the parameter of the arc.
  - The ending point of the arc.
  - The direction of the arc, which is the tangent that passes through the start point. You should input the angle of the tangent.
  - The distance between the starting point and the ending point, which is called the Length of Chord.
  - The center point of the arc.
  - The radius.
  - The angle between Start-Center-End, which is called the Included Angle.
- See the following illustration:

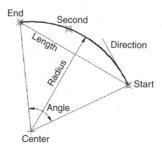

- If you provide three of these eight, AutoCAD will be able to draw an arc, but you cannot provide just any three. The combination of the information needed can be found in the **Home** tab, using the **Draw** panel, while clicking the arrow near the **Arc** button to see all the available methods:

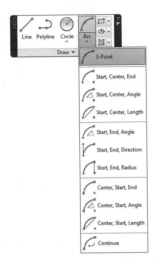

- As you can see the Start point is always one of the required pieces of information. You should think CCW when specifying points. On the other hand if you want to work clockwise, simply hold the [Ctrl] key and it will change.

2.8 OSNAPS RELATED TO CIRCLE AND ARC

- Some of OSNAPs relate to circles and arcs are:
  - **Center**: To catch the Center of arc or circle
  - **Quadrant**: To catch the Quadrant of arc, or circle
  - **Tangent**: To catch the Tangent of arc, circle
- See the following illustrations:

2.9 USING OTRACK WITH OSNAP

- Sometimes OSNAP alone is not enough to specify desired points, especially if you need complex points. To solve this problem in the past we used to draw dummy objects to help us specify complex points. For example, when we wanted to specify the center of the circle at the center of a rectangle we used to a draw line from the midpoints of the two vertical lines, and the same for the horizontal lines. But since the introduction of Object Snap Tracking, or OTRACK, in AutoCAD 2000, drawing dummy objects has no longer been needed. OTRACK depends on active OSNAP modes, which means if you want to use the midpoint with OTRACK, you have to switch the midpoint on first. To activate OTRACK go to the status bar and click the **Object Snap Tracking** button on:

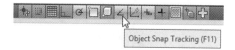

- The procedure is very simple:
  - Using OSNAP, go to the desired point and hover over it for couple of seconds (don't click), then move to the right or left (also up and down depending on the next point), and you will see an infinite line extending in both directions (this line will be horizontal or vertical depending on your movement).
  - If you want to use a single point to specify your desired point, move in the desired direction, type in the desired distance, and press [Enter].
  - If you need two points, go the next point, and stay for a couple of seconds, then move in the desired direction. Another infinite line will appear. Go to the intersection point of the two infinite lines, and that will be your point.

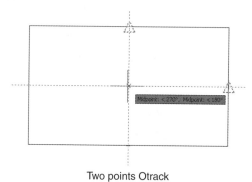

Two points Otrack

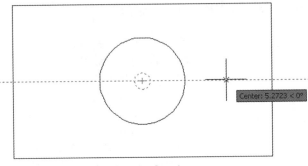

Single point Otrack

- The Polar command is a major help here as well. Let's go back to the same dialog box:

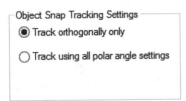

- Under **Object Snap Tracking Settings**, there are two choices:
  - Track orthogonally only (default option)
  - Track using all polar tracking settings
- This means you can use the current polar angles (increment and additional angles) to specify points using OTRACK.

 - To deactivate an OTRACK point, hover at the same point again for a couple of seconds and it will be deactivated.

PRACTICE 2-4

 Drawing Using OSNAP and OTRACK

1. Start AutoCAD 2014.
2. Open **Practice 2-4.dwg**.

3. Using the proper **OSNAP**, create the four arcs as shown below:

4. Create the two circles as shown below (radius = 1):

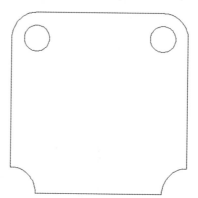

5. Using **OSNAP** and **OTRACK** (using two points), draw the circle at the center of the shape (radius = 3.0):

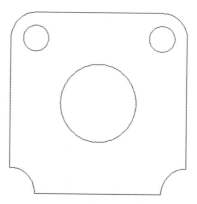

6. Using OSNAP and OTRACK (one point), draw the two circles at the right and left (distance center-to-center = 5.0 and radius = 0.5):

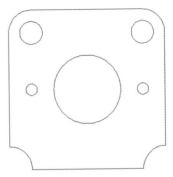

7. Change the increment angle in the Polar Tracking dialog box to 45. Make sure that **Track using all polar angle settings** is on and draw a circle (radius = 0.5), its center specified using the OSNAP and OTRACK and Polar Tracking as shown below:

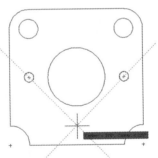

8. Use the same procedure to draw a circle at the top to end up with the final shape as shown here:

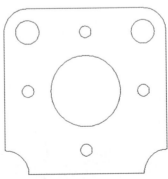

9. Save and close the file.

2.10 DRAWING LINES AND ARCS USING THE POLYLINE COMMAND

- The Polyline command allows you to do all or any of the following:
  - Draw both line segments and arc segments.
  - Draw a single object with the same command rather than drawing segments of lines and arcs, as with the Line and Arc commands.
  - Draw lines and arcs with starting and ending widths.
- To use the command, go to the **Home** tab, locate the **Draw** panel, then select the **Polyline** button:

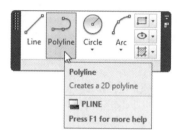

- The following prompts will appear:

```
Specify start point:
Current line-width is 1.0000
Specify next point or [Arc/Halfwidth/Length/Undo/ Width]:
```

- AutoCAD will ask you to specify the first point, and when you do, AutoCAD will report to you the current line-width. If you like it, continue specifying points using the same method we learned with the Line command; if not, change the width as a first step by typing the letter **w**, or right-clicking and selecting the **Width** option. You will see the following prompts:

```
Specify starting width <1.0000>:
Specify ending width <1.0000>:
```

- Specify the starting width, press [Enter], and then specify the ending width. The next time you use the same file, AutoCAD will report these values to you when you issue the Polyline command. Halfwidth is the same, but instead of specifying the full width, you specify the halfwidth.
- The Undo and Close options are identical to the ones in the Line command.

- **Length** allows you to specify the length of the line using the angle of the last segment.
- **Arc** l allows you to draw an arc attached to the line segment. You will see the following prompt:

```
Specify endpoint of arc or [Angle/CEnter/CLose/ Direc-
tion/ Halfwidth/Line/Radius/Second pt/Undo/Width]:
```

- Arc will be attached to the last segment of line, or will be the first object in a Polyline command. Using either method, the first point of the arc is already known, so we need 2 more pieces of information. AutoCAD will make an assumption (which you can reject) that the angle of the last line segment will be considered the direction (tangent) of the arc. If you accept this assumption, you should specify the endpoint. If not, choose from the following to specify the second piece of information:
    - The Angle of the Arc
    - The Center point of the arc
    - Another Direction to the arc
    - The Radius of the arc
    - The Second point which can be any point on the parameter of the arc
- Based on the information selected as the second point, AutoCAD will ask you to supply the third piece of information.

2.11 CONVERTING POLYLINES TO LINES AND ARCS AND VICE-VERSA

- This is a very essential technique that allows you to convert any polyline to lines and arcs and convert lines and arcs to polylines.

2.11.1 Converting Polylines to Lines and Arcs

- The **Explode** command will explode a polyline into lines and arcs. To issue this command, go to the **Home** tab, locate the **Modify** panel, then select the **Explode** button:

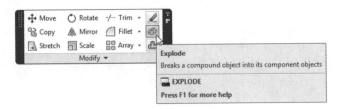

- AutoCAD will show the following prompt:

```
Select objects:
```

- Select the desired polylines and press [Enter] when done. The new shape will have lines and arcs.

2.11.2 Joining Lines and Arcs to Form a Polyline

- Here, we will discuss an option called **Join** within a command called **Edit Polyline**. To issue this command, go to the **Home** tab, the locate **Modify** panel, then select the **Edit Polyline** button:

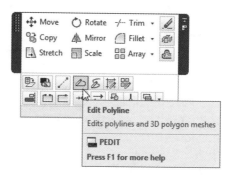

- You will see the following prompts:

```
Select polyline or [Multiple]:
Object selected is not a polyline
Do you want to turn it into one? <Y>
Enter  an  option  [Close/Join/Width/Edit  vertex/Fit/
Spline/Decurve/Ltype gen/Reverse/Undo]: J
```

- Start by selecting one of the lines or arcs you want to convert. AutoCAD will respond by telling you that the selected object is not a polyline! and giving you the option to convert this specific line or arc to a polyline. If you accept this, options will appear, and one of these options will be **Join**. Select the **Join** option, then select the rest of the lines and arcs. At the end, press [Enter] twice. The objects will be converted to a polyline.

PRACTICE 2-5

Drawing Polylines and Converting Them

1. Start AutoCAD 2014.
2. Open **Practice 2-5.dwg**.
3. Draw the following polyline using a start point of 18.5 and width = 0.1.

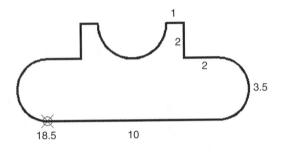

4. Then explode the polyline. As evidence, the width will disappear.
5. After exploding, the objects will be lines and arcs.
6. Save and close.

2.12 USING SNAP AND GRID TO SPECIFY POINTS ACCURATELY

- Snap and Grid is another method of specifying points accurately in the X,Y plane.
- Using the mouse alone (the default method) is not accurate, so we can't depend on it to specify points. We need to control its movement, which is the sole function of Snap. Snap can control the mouse and make it jump in the X and Y directions in exact distances.
- Grid by itself is *not* accurate tool, but it will complement Snap function. It will show horizontal and vertical lines replicating the drawing sheets. In order to turn on/off the **Snap**, use the following button in the status bar:

- In order to turn on/off the **Grid** tool, use the following button on the status bar:

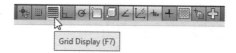

- Most of the time, using both tools at the same time will not be enough, but you can modify the settings to fit your needs. Right-click one of the two buttons and the following shortcut menu will appear:

- Select the **Settings** option and the Drafting Settings dialog box will appear:

- Input the Snap X Spacing and Snap Y Spacing (by default they are equal). Switch off the checkbox to make them unequal. Do the same for the Grid Spacing in X and Y. If you want Grid to follow Snap, set the Grid spacing to zeros. In Grid there are major and minor lines; set the major line frequency.

- Set if you want to see the Grid in dots and where (2D model space, Block editor, or Sheet/layout).
- Grid behavior is for 3D only.
- Specify the type of Snap: Grid Snap or Polar Snap. Polar Snap will work with Polar Tracking and allow you to specify exact distances in angles such as 30, 45, 60, etc. You can also use the following function keys to turn on/off both Snap and Grid:
  - F9 = Snap on/off
  - F7 = Grid on/off

2.13 USING POLAR SNAP

- While using the Snap command, the mouse will use the increment distance to specify exact distance only in horizontal and vertical directions. While going diagonal distances, Snap will not help. To solve this problem AutoCAD introduced Polar Snap, which will specify exact increments using all angles specified in the Polar Tracking dialog box. To activate Polar Snap, AutoCAD will switch off normal Snap (called Grid Snap). To do that go to the Snap button (or Grid button) at the status bar, and select the Settings option. You will see the following dialog box:

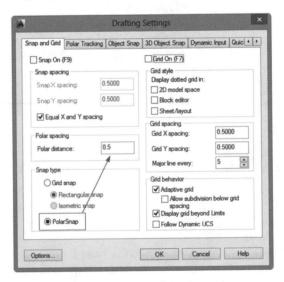

- Select the Polar Snap button, then specify the Polar distance as shown in the above illustration.

PRACTICE 2-6

Snap and Grid

1. Start AutoCAD 2014.
2. Open **Practice 2-6.dwg**.
3. Change Polar Tracking to use angle = 45.
4. Change Snap to Polar Snap and set the distance to 0.5.
5. Draw the following shape, starting from point 22,5, all segment lengths 6.5, and angle 45:

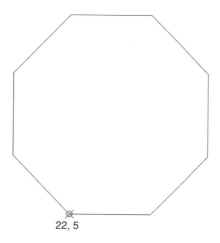

22, 5

6. Save and close the file.

NOTES:

CHAPTER REVIEW

1. The Polyline command is different from the Line command because:
   a. It will produce lines and arcs.
   b. All segments drawn using the same command are considered a single object.
   c. You can specify a starting and ending width.
   d. All of the above.
2. Perpendicular, Tangent, and Endpoint are some _____ available in AutoCAD.
3. OTRACK can work by itself.
   a. True
   b. False
4. Which one of the following is *not* one of the eight pieces of information AutoCAD needs to draw an arc?
   a. End point of the arc
   b. Midpoint of the arc
   c. Center point of the arc
   d. Angle
5. To convert lines and arcs from a polyline, use the _____ command.
6. If you want Grid to follow the Snap settings, set it up to be _____ in both X and Y.
7. Using the Polar Tracking dialog box, you can only track the orthogonal angles.
   a. True
   b. False
8. The_____ is always required to draw an arc.
9. To convert lines and arcs to polylines use the:
   a. Polyline Edit, Convert option
   b. Polyline Edit, Union option
   c. Polyline Edit, Join option
   d. None of the above.

CHAPTER REVIEW ANSWERS

1. d
3. b
5. Explode
7. b
9. c

3

MODIFYING COMMANDS PART I

In This Chapter

◇ Different methods for selecting objects and selection cycling
◇ How to erase objects
◇ How to move and copy objects
◇ How to rotate and scale objects
◇ How to mirror, stretch, and break objects
◇ How to lengthen, and join objects
◇ How to use grips for editing

3.1 HOW TO SELECT OBJECTS IN AUTOCAD

■ In order to use any of the modification commands we will discuss in this chapter, you have to select the desired objects. Once you issue any of the modifying commands, the following prompt will appear:

```
Select objects:
```

■ The cursor will change to a pick box. At this prompt, you can work without typing anything, or you can type a few letters to activate a certain mode

■ Without typing any letter, you can do the following things:
  • Select objects by clicking them using the pick box one-by-one.
  • Without selecting any object, click and move to the right to start Window mode, which allows you to select all objects contained fully inside the window.
  • Without selecting any object, click and move to the left to start Crossing mode, which allows you to select all objects contained fully inside the crossing or touched (crossed) by it.

- See the following two examples:
- Window Example:

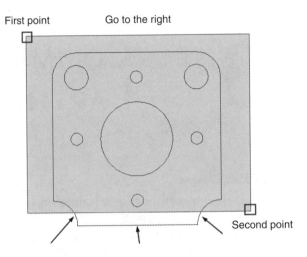

- All objects will be selected except the three objects with an arrow pointing to them. Why? Because either they are not fully contained inside the window, or because they are not contained at all.
- Crossing Example:

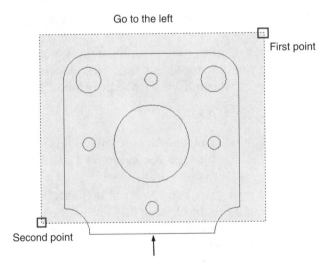

- All objects will be selected except the horizontal line with an arrow pointing to it. Why? Because it was neither contained nor crossed.

- Another way to use the Select objects prompt is to type a few letters to activate a certain mode. These letters are discussed as follows.

3.1.1 Window Mode (W)

- At the command prompt, type **W**, and you will switch the selecting mode to **Window**, which will be available whether you go to the right or to the left.

3.1.2 Crossing Mode (C)

- At the command prompt, type **C**, and you will switch the selecting mode to **Crossing**, which will be available whether you go to the right or to the left.

3.1.3 Window Polygon Mode (WP)

- WP mode allows you to specify a non-rectangular window by specifying points in any fashion you like. After typing **WP** and pressing [Enter], the following prompts will appear:

```
First polygon point:
Specify endpoint of line or [Undo]:
Specify endpoint of line or [Undo]:
```

- Press [Enter] to end WP mode. WP is just like W; it needs to contain the object fully in order to select it.
- See the following example:

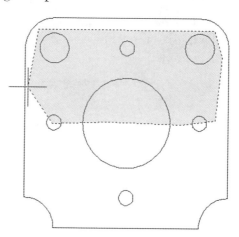

- Only four circles will be selected, and they are fully contained inside the WP.

3.1.4 Crossing Polygon Mode (CP)

- Since there is W and WP, it is obvious that there is also **C** and **CP**. Crossing Polygon mode is just like WP mode; it allows you to specify a non-rectangular shape to contain and cross objects. See the following illustration:

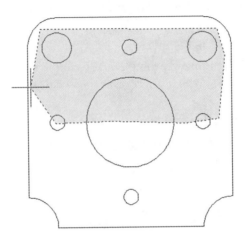

- Three circles will be selected because they are fully contained inside the CP, and three more will be selected because they are crossed by the CP.

3.1.5 Fence Mode (F)

- **Fence** mode allows you to select multiple objects by crossing (touching) them. The lines of the fence can cross, contrary to WP and CP. See the following example:

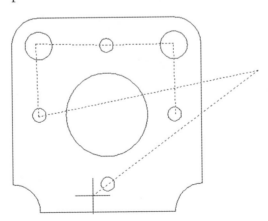

- All the circles are selected because they are touched by the fence. The vertical line at the right will be selected as well. You can see that the fence lines are crossed, which is 100% acceptable in AutoCAD.

3.1.6 Last (L), Previous (P), and All Modes

- You can also select objects using the following modes:
  - **Last (L):** To select the last object drawn.
  - **Previous (P):** To select the last selected objects.
  - **All:** To select all objects in the drawing.
- NOTE ▸ To deselect objects, hold the [Shift] key on the keyboard and then click the objects. While you are in this mode you can use Window and Crossing modes.

3.1.7 Other Methods to Select Objects

- There are two ways to use Modifying commands. You can issue the command, then select objects, or you can select objects, then issue the command. This technique is called the **Noun/Verb** technique. Without issuing any command, you can:
  - Select a single object by clicking it.
  - Click an empty space and go the right to get to **Window** mode.
  - Click an empty space and go the left to get to **Crossing** mode.
  - Click an empty space, then type W to get to **Window Polygon** mode.
  - Click an empty space, then type C to get to **Crossing Polygon** mode.
  - Click an empty space, then type F to get to **Fence** mode.
- When you select objects, you can go to the **Home** tab, locate the **Modify** panel, and issue the desired command, or you can right-click, and you will get a shortcut menu that contains five modifying commands: Erase, Move, Copy Selection, Scale, and Rotate, as shown below:

- By default, **Noun/Verb** is on, but to change it, take the following steps:
  - Go to the **Application Menu** and select **Options** button.
  - Select the **Selection** tab.
  - Under **Selection modes**, make sure that **Noun/Verb** selection is on.

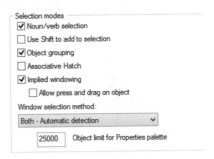

- Use this dialog box as well to do the following:
  - Turn on **Allow press and drag on object** to create a window or crossing even if the cursor is not over a clean spot for picking.
  - Under **Window selection method**, make sure that **Both Automatic detection** is selected so if you click then release the mouse, or you click and drag, both methods will be accepted.

3.2 SELECTION CYCLING

- While you are drafting using AutoCAD you may unintentionally draft objects over each other. Or, you may click on an object using a point that is shared by other objects. This issue used to be a problem in the past, but not anymore. The **Selection Cycling** feature in AutoCAD will notify you if your click touches more than one object and will give you the chance to choose the desired one. To activate **Selection Cycling** (by default it is active), go to the status bar and click the button as shown:

- After activating it and clicking on one object, you may see the following:

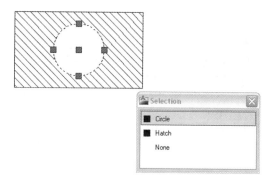

- The small window tells you that there are two possible objects, Circle and Hatch, and the Circle is selected. Using this window, if you move the mouse to Hatch, you will get the following:

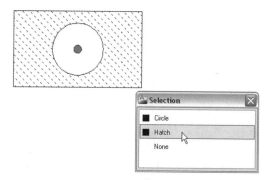

3.3 ERASE COMMAND

- **The Erase** command will delete any object you select. To issue the command, go to the **Home** tab, locate the **Modify** panel, and then select the **Erase** button:

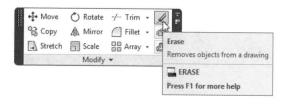

- You will see the following prompt in the Command window:

  `Select Objects:`

- Select the desired objects. The Select Objects prompt is repetitive, so you will always need to end it by pressing [Enter], or by right-clicking.
- You can also erase using other methods:
  - Click on the desired object(s) and then press the [Del] key on the keyboard.
  - Click on the desired object(s) and then right-click and a shortcut menu will appear; select **Erase**:

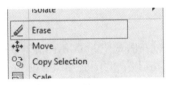

PRACTICE 3-1

Selecting Objects and the Erase Command

1. Start AutoCAD 2014.
2. Open **Practice 3-1.dwg**.
3. Start the Erase command (for all steps, after you finish selecting, press [Enter], then Undo what you did).
4. Without typing any letter, find an empty space at the left, click (release your finger, don't hold the mouse button down) and go the right. Try to get the following result:

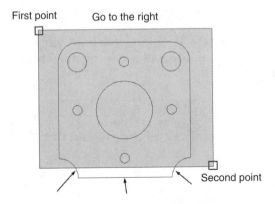

5. What was selected? Did you expect this result?
6. Do the same steps to simulate the following:

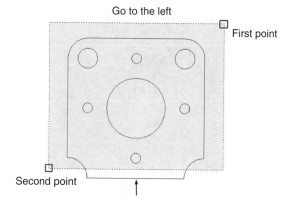

7. What was selected and why? _____
8. Try to simulate the following using WP:

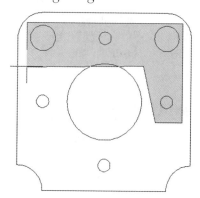

9. Try to simulate the following using CP:

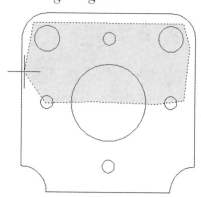

10. Try to simulate the following using Fence:

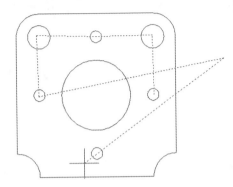

11. Without issuing the Erase command, select objects, then press [Del] on the keyboard to delete them. Undo what you did.
12. Without issuing the Erase command, select objects, then right-click and select the Erase command from the menu. Undo what you did.
13. Start the Erase command, select objects, then hold [Shift] to deselect some of the selected objects, and press [Enter]. Undo what you did.
14. Close the file without saving.

3.4 MOVE COMMAND

- This command will move objects from one place to another in the drawing. To issue this command, go to the **Home** tab, locate the **Modify** panel, then select the **Move** button:

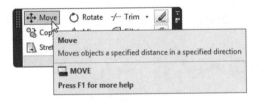

- The following prompts will be shown:

```
Select objects:
Specify base point or [Displacement] <Displacement>:
Specify second point or <use first point as displacement>:
```

- The first step is to select objects, and then you should select the base point. The base point is the point that will represent the objects; it will move a distance at an angle, and objects will follow. The main objective of the base point is accuracy. The last prompt asks you to specify the second point, or the destination of your movement.
- See the following illustration:

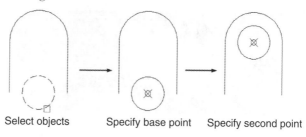

Select objects   Specify base point   Specify second point

3.4.1 Nudge Functionality

- This function is very simple and will enable you to make an orthogonal move for selected objects. All you have to do is select objects, hold the [Ctrl] key on the keyboard, then use the four arrows on the keyboard, you will see objects move in the desired direction.

PRACTICE 3-2

Moving Objects

1. Start AutoCAD 2014.
2. Open **Practice 3-2.dwg**.
3. Using the Move command, move the three circles and the rectangle to the correct places, so you get the following result (you will need OSNAP and OTRACK to move the rectangle accurately):

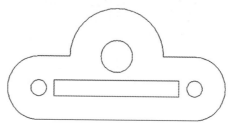

4. Save and close the file.

3.5 COPY COMMAND

- This command will copy objects. To issue this command, go to the **Home** tab, locate the **Modify** panel, then select the **Copy** button:

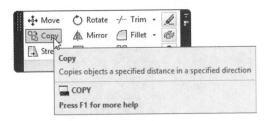

- You will see the following prompts:

```
Select objects:
Current settings: Copy mode = Multiple
Specify base point or [Displacement/mOde] <Displacement>:
Specify second point or [Array] <use first point as
displacement>:
Specify second point or [Array/Exit/Undo] <Exit>:
Specify second point or [Array/Exit/Undo] <Exit>:
```

- After you select the desired objects, AutoCAD will give you the current mode, in this case, Multiple. This mode allows you to create several copies in the same command. The other mode is Single copy. The first prompt asks you to specify the base point. AutoCAD will then ask you to specify the second point to complete a single copy process, then repeats the prompt to create another one, and so on. There are three options you can use:
  - Undo, to undo the last copy.
  - Exit, to end the command.
  - Array, which allows you to create an array of the same object using distance and angle. When you select the array option you will see the following prompts:

```
Enter number of items to array:
Specify second point or [Fit]:
```

- The first prompt is to input the number of items to array (including the original object); you will then specify the distance between the objects.

You can use the **Fit** option to specify the total distance, and AutoCAD will equally divide the distance over the number of objects.

- See the following example of multiple copies:

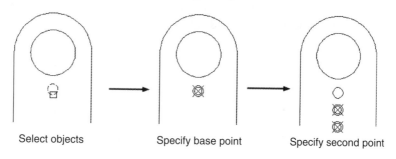

| Select objects | Specify base point | Specify second point |

PRACTICE 3-3

 Copying Objects

1. Start AutoCAD 2014.
2. Open **Practice 3-3.dwg**.
3. Copy the door using Multiple copying, and copy the toilet using the Array option (use the midpoint OSNAP for the toilet) to achieve the following:

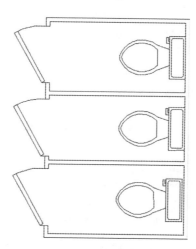

4. Save and close the file.

3.6 ROTATE COMMAND

- This command will rotate objects around the base point, using a rotation angle or reference. To issue the command, go to the **Home** tab, locate the **Modify** panel, and select the **Rotate** button:

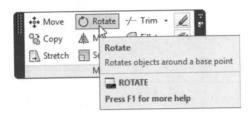

- You will see the following prompts:

```
Current positive angle in UCS:
ANGDIR=counterclockwise ANGBASE=0
Select objects:
Select objects:
Specify base point:
Specify rotation angle or [Copy/Reference] <0>:
```

- The first message gives you the current angle direction and the angle base value. The base point here is the rotation point, which all the selected objects will rotate around.
- Use the **Copy** option to make a copy of the selected objects and then rotate them.
- See the following example:

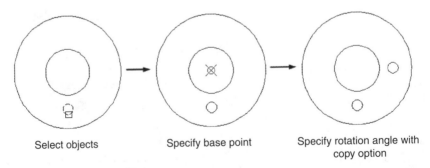

Select objects   Specify base point   Specify rotation angle with
                                                     copy option

3.6.1 Reference Option

- This option is very helpful if you don't know the rotation angle. You specify two points to indicate the current angle, and two points to input the new angle. You will see the following prompts:

```
Specify the reference angle <0>:
Specify second point:
Specify the new angle or [Points] <0>:
```

- You can input the angle by typing. If you want to input angles using two points, the first point you will choose will be for both angles; the current and the new. See the following illustration:

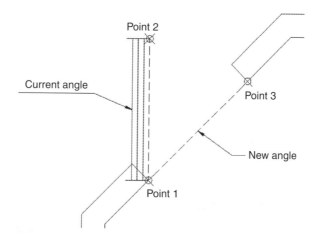

PRACTICE 3-4

  Rotating Objects

1. Start AutoCAD 2014.
2. Open **Practice 3-4.dwg**.
3. Rotate the lower window using Angle.
4. Rotate the upper window using Reference.

5. Rotate the chair with Copy mode to get the following result:

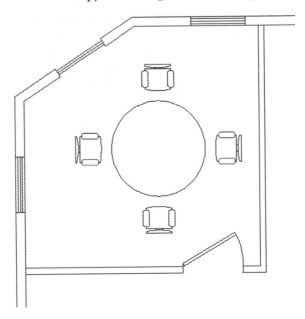

6. Save and close the file.

3.7 SCALE COMMAND

- This command will help you create larger or smaller objects using a scale factor, or reference. To issue this command, go to the **Home** tab, locate the **Modify** panel, then select the **Scale** button:

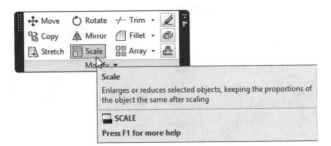

- You will see the following prompts:

```
Select objects:
Specify base point:
Specify scale factor or [Copy/Reference] <1.0000>:
```

- The base point here is the scaling point; all the selected objects will be larger or smaller relative to it. Use the **Copy** option to make a copy of the selected objects and then scale them. See the following example:

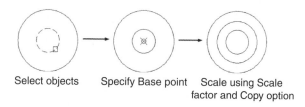

Select objects   Specify Base point   Scale using Scale factor and Copy option

3.7.1 Reference Option

- This option is very handy if you don't know the scaling factor as a number; you can specify two points to indicate the current length, and two points to input the new length. You will see the following prompts:

```
Specify reference length <0'-1">:
Specify second point:
Specify new length or [Points] <0'-1">:
```

- You can input length by typing. If you want to input lengths using two points, the first point you will choose will be for both lengths; the current and the new. See the following illustration:

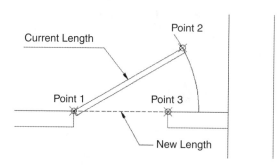

PRACTICE 3-5

Scaling Objects

1. Start AutoCAD 2014.
2. Open **Practice 3-5.dwg**.
3. Scale the toilet by scale factor = 0.9 using the mid-point of the wall.
4. Scale the sink by scale factor = 1.2 using the quadrant of the sink.
5. Scale the door using the Reference option to fit in the door opening.
6. You will get the following:

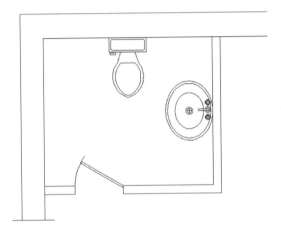

7. Save and close the file.

3.8 MIRROR COMMAND

- This command will help you create a mirror image of the selected objects using a mirror line. To issue this command, go to the **Home** tab, locate the **Modify** panel, then select the **Mirror** button:

- You will see the following prompts:

```
Select objects:
Specify first point of mirror line:
Specify second point of mirror line:
Erase source objects? [Yes/No] <N>:
```

- After selecting objects, specify the mirror line by specifying two points (you don't need to draw a line to be able to specify these two points). The last prompt asks you to keep or delete the original objects.
- Text can be part of the selection set to be mirrored. You can tell Auto-CAD what to do with it (copying or mirroring) by using system variable MIRRTEXT. To issue this command, type it in the Command window, and you will see the following:

```
Enter new value for MIRRTEXT <0>:
```

- You can input either 1 (which means mirror the text just like the other objects) or 0 (which means copy the text).
- See the following example:

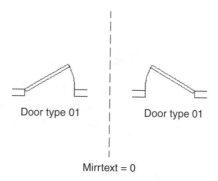

Door type 01   Door type 01

Mirrtext = 0

PRACTICE 3-6

 Mirroring Objects

1. Start AutoCAD 2014.
2. Open **Practice 3-6.dwg**.
3. Mirror the entrance door to open inside rather than outside, keeping the text as is.

4. Mirror all windows at the right to the left.
5. Mirror the furniture of the room at the right to the room in the left.
6. Mirror the two windows at the top (to the one at the right and the one at the middle) to the lower wall.
7. You should get the following result:

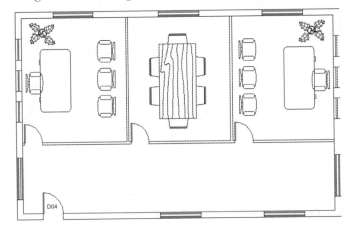

8. Save and close.

3.9 STRETCH COMMAND

- This command allows you to change the length of selected objects by stretching using distance and angle. To issue this command, go to the **Home** tab, locate the **Modify** panel, and select the **Stretch** button:

- You will see the following prompts:

```
Select objects to stretch by crossing-window or crossing-
polygon...
Select objects:
```

```
Specify base point or [Displacement] <Displacement>:
Specify second point or <use first point as displacement>:
```

- The Stretch command is different than other modifying commands we have learned about because it asks you to select the objects desired using C or CP modes. Why? Because Stretch will utilize both features of C and CP, containing and crossing. All objects contained fully inside the C or CP will be moving, while objects crossed will be stretched either by increasing or decreasing the length. You should then specify the base point (the same principle of the Move and Copy commands), and finally, specify the second point. See the following example:

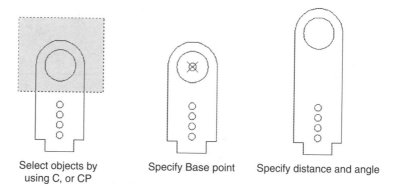

Select objects by   Specify Base point   Specify distance and angle
using C, or CP

PRACTICE 3-7

Stretching Objects

1. Start AutoCAD 2014.
2. Open **Practice 3-7.dwg**.
3. The vertical distance of the three rooms is not correct; it should be 1'-0" more. Use the Stretch command to fix it.
4. The door of the room at the right is positioned incorrectly; stretch it to the right, for 2'-5".
5. The door of the room at the left is also positioned incorrectly; stretch it to the right to the small line indicating the new position.

6. You should have the following:

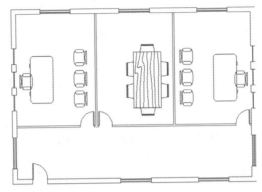

7. Save and close the file.

PRACTICE 3-8

 Stretching Objects

1. Start AutoCAD 2014.
2. Open **Practice 3-8.dwg**.
3. Stretch the upper part to look like the lower part by using distance = 1.00 and CP.
4. You should have the following:

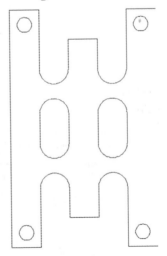

5. Save and close the file.

3.10 LENGTHENING OBJECTS

- Using this command you can add length or subtract length from objects using different methods. To issue this command, go to the **Home** tab, locate the **Modify** panel, and then select the **Lengthen** button:

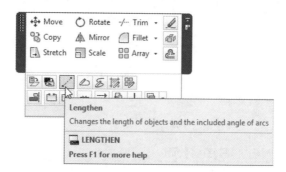

- You will see the following prompt:

```
Select an object or [DElta/Percent/Total/DYnamic]:
```

- If you click an object, AutoCAD will report the current length using the current unit. There are multiple methods to lengthen (or shorten) objects in AutoCAD.
- The first method is called **Delta**, and it allows you to add (remove) to (from) the current length. A positive value means adding, and a negative value means subtracting.
- This is the prompt you will see if you choose the Delta option:

```
Enter delta length or [Angle] <0.0000>:
```

- The second method is called **Percentage**, and it allows you to add (remove) to (from) the length by specifying a percentage of the current length. To add more length input a value more than 100; to remove length input a value less than 100.
- You will see the following prompt if you choose the Percentage option:

```
Enter percentage length <100.0000>:
```

- The third method is **Total**, which allows you to input a new total length of the object. This means the object will add length if the new value is greater than the current value, and vice-versa.

- You will see this prompt if you choose the Total option:

  ```
  Specify total length or [Angle] <1.0000)>:
  ```

- The fourth method is **Dynamic**, and it allows you to increase/decrease the length dynamically using the mouse. You will see the following prompt:

  ```
  Specify new end point:
  ```

- You lengthen/shorten a single object using a single method per command.

3.11 JOINING OBJECTS

- The Join command is helpful because it will join lines to other lines, arcs to arcs, and polylines to polylines.
- To issue this command, go to the **Home** tab, locate the **Modify** panel, and select the **Join** button:

- AutoCAD will show the following prompts:

  ```
  Select source object or multiple objects to join at once:
  Select lines to join to source:
  Select lines to join to source:
  1 line joined to source
  ```

- The above prompts are for joining lines, and they may differ for arcs, and polylines. There are some conditions for the joining to succeed:
  - You can join lines, arcs, and polylines to form a single polyline, but they should be connected to the ends of each other.
  - If the lines, arc, and polylines are not connected, each connected group will be considered a single polyline.
  - If you want to join lines to form a line they should be always collinear.
  - If you want to join arcs to form a single arc, they should have the same center point.
  - While joining arcs, there is a special prompt asking if you want to create a circle (this is the only command that allows you to create a circle from an arc).

PRACTICE 3-9

Lengthen and Joining Objects

1. Start AutoCAD 2014.
2. Open **Practice 3-9.dwg**.
3. Using the Lengthen command make the total length of the line at the upper left = 1.75".
4. Using the Lengthen command make the arc length at the right side of the shape 200% of the current length.
5. Using the Join command join the arc at the top and the line at its right.
6. Using the Join command convert the two arcs at the top and at the bottom to be a full circle.
7. Using the Join command join all objects to the polyline.
8. You should get the following shape:

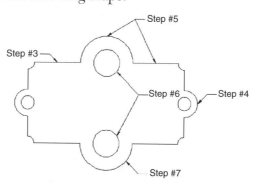

9. Save and close the file.

3.12 USING GRIPS TO EDIT OBJECTS

- **Grips** are blue squares and rectangles that appear on objects and enable you to modify them using five modifying commands using a grip as the base point. It is a clever tool for performing modifying tasks faster without compromising the accuracy of the conventional modifying commands discussed earlier.
- Depending on the type of object, grips will appear in different places. See the following illustration:

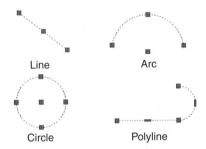

- When you click an object, it will be selected, and you can issue any of the modifying commands we have discussed. But if you click any of the blue squares (grips) the object will turn red, which means it will become the active base point (hot) for the following five modifying commands:
  - Stretch
  - Move
  - Rotate
  - Scale
  - Mirror
- Sometimes you can use the Lengthen command (discussed in the next chapter) depending on which grip was clicked and which object was selected. You can see these five commands by right-clicking:

- These commands have one thing in common, the base point, except for Mirror. So why is Mirror in the list? And why isn't Copy among the commands in the list? To answer the first question, AutoCAD considers the first point of the mirror line as the base point, so it is included in the list. To answer the second question, Copy here is a mode and not a command, which means it will work with all five commands.
- Holding [Shift] while selecting the grip will enable you to select more than one base point.
- Holding [Ctrl] while specifying the second point (Move and Stretch), rotation angle (Rotate), scale factor (Scale), or second point of the mirror line (Mirror) allows AutoCAD to remember the last input and repeat it graphically.
- The Base Point option allows you to select another base point other than the grip selected.
- Let's look at an example. Select the following shapes (two polylines):

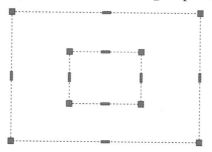

- Click the upper-right corner to make it hot, then right-click to access the menu and select the Rotate command:

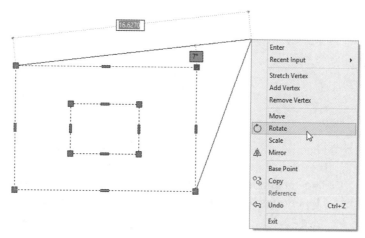

- Now you can rotate the two shapes around the grip (considered your base point); right-click again and select the Base Point option to select another base point, which will be the center of the two rectangles (using OSNAP and OTRACK):

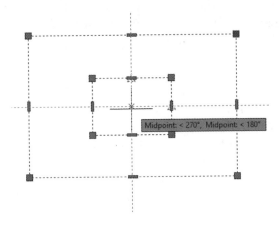

- Right-click again and select the Copy option to copy while rotating. Using Polar Tracking, specify an angle of 90 and then press [Esc]. You will get the following result:

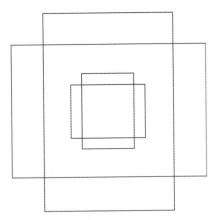

- Specific objects such as polylines will show more than the blue squares; they will show rectangles at the midpoint of each line and arc segment. The rectangles at the midpoint of the polyline have several functions. Depending on whether it is a line segment or arc segment, AutoCAD will show a different menu. Go to the grip and hover over it for a second (don't click) and you will see something like the following:

- As you can see, you can Stretch the selected segment; Add a vertex, or converting a line to an arc, or converting an arc to line.
- When you are done with grips click [Esc] once or twice depending on the situation you are in, and this will end the grips mode.

3.13 GRIPS AND DYNAMIC INPUT

- If you stayed on one of the grips (without clicking), Dynamic Input along with grips will help you get information about the selected objects based on their type. See the following examples:
- Using the Line command and one of the endpoints, you will see the length and angle with the east, along with two commands, Stretch and Lengthen:

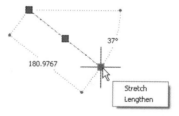

- The two connected lines will show two lengths and two angles:

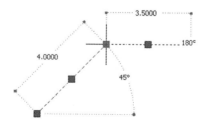

- The midpoint of an arc will show the radius and included angle, along with two commands, Stretch and Radius:

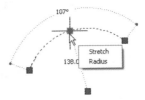

- However, the endpoints of an arc will show the radius and angle with the East, along with the Stretch and Lengthen commands:

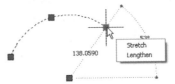

- As for a circle, you will only see the radius:

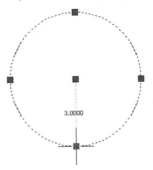

- The endpoint of a segment of a polyline will show the length of the shared lines, along with a shortcut menu that includes options such as Stretch Vertex, Add Vertex, and Remove Vertex:

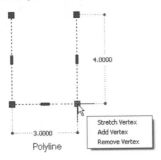

3.14 GRIPS AND PERPENDICULAR AND TANGENT OSNAPS

- Using grips you can specify perpendicular and tangent OSNAPs, keeping in mind that those two settings are turned on when running OSNAP. See the following two examples:
- Tangent Example:

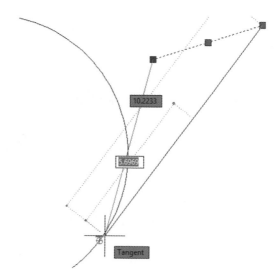

- Perpendicular Example:

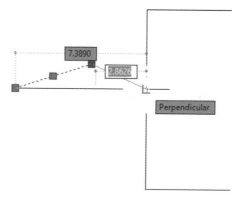

PRACTICE 3-10

Using Grips to Edit Objects

1. Start AutoCAD 2014.
2. Open **Practice 3-10.dwg** (you should complete this practice using grips techniques only).
3. Select the large circle, make the center hot, and scale it with copying, with 1.2 as the scale factor.
4. Mirror the three lines at the right to the other side (you should use another base point, along with Copy mode).
5. You should have the following shape:

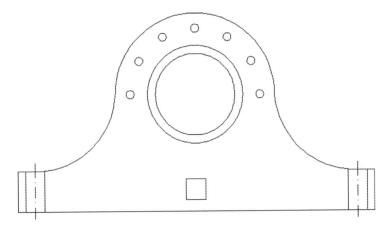

6. Select the polyline below the two big circles. Convert the top horizontal line to an arc by moving up a distance = 0.2 (this is not the new radius).
7. Using the rectangle grip at the middle of the lower horizontal line of the polyline stretch the line downward by distance = 0.1.
8. Using grips and Dynamic Input what is the horizontal distance and the vertical distance of the polyline? _____, _____
9. Press [Esc] to clear grips, then re-select the polyline again, select one of the grips (any one), and then right-click and select the Move command, then right-click again, and select Copy mode (make sure Polar Tracking is on to help you get exact angles). Now move to the right, type 1 as a distance, but before you press [Enter] hold the [Ctrl] key so AutoCAD

will remember this distance. Now while you are still holding the [Ctrl] key, make three copies to the right and three copies to the left.

10. The shape should look like the following:

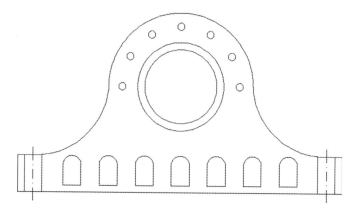

11. Save and close the file.

NOTES:

CHAPTER REVIEW

1. Which one of the following is *not* true about the Reference option?
   a. You can use it with Rotate and Scale.
   b. In Rotate, you have to input to specify the current and the new angle by specifying four points.
   c. In Scale you have to specify two distances, the current and the new.
   d. You can't use it in the Stretch command.
2. The Stretch command will ask for something that other modifying commands will not.
   a. True
   b. False
3. If you want to select the last selected set, type _____ at the Command window.
4. In the Mirror command, you should draw a line to act as the mirror line.
   a. True
   b. False
5. While using grips holding _____ while selecting the grip will enable you to select more than one base point.
6. _____ will help you control the outcome of text in the Mirror command
7. In grips, there are _____ modifying commands in the right-click menu.
   a. Seven
   b. Six
   c. Five
   d. Four
8. While using the Lengthen command, the option to double the current length without knowing it is:
   a. Delta
   b. Total
   c. Percent
   d. Dynamic

CHAPTER REVIEW ANSWERS

1. b
3. P
5. [Shift]
7. five

Chapter 4

MODIFYING COMMANDS PART II

In This Chapter

◇ How to offset objects
◇ How to fillet and chamfer objects
◇ How to trim and extend objects
◇ How to array objects using three different methods
◇ How to break objects

4.1 INTRODUCTION

- The commands discussed in this chapter are modifying commands with special abilities. They can build over the shapes you draw. Each one of them has a unique function; two of them can create objects (Offset and Array), and the others can change an existing object's shape (the Fillet, Chamfer, Trim, Extend, Lengthen, and Join commands).
- Here is a brief description of each command:
  - **Offset** command: used to create copies of an object parallel to the original.
  - **Fillet** command: used to create a neat intersection between two objects either by extending/trimming lines or using arcs.
  - **Chamfer** command: used to create a neat intersection but only for lines. The neat intersection is created by extending/trimming the two lines or by creating a new line showing the chamfered edge.
  - **Trim** command: used to trim objects using other objects as cutting edges.
  - **Extend** command: used to extend objects using other objects as boundary edges.

- **Array** command: used to create objects using three different methods: rectangular, circular, and using a path.
- **Break** command: used to break an object into two objects by specifying two points and deleting the portion between them.

4.2 OFFSETTING OBJECTS

- The Offset command will create copies of an object parallel to the original. The new object will have the same properties as the original object. You can offset using offset distance, or using a point the new object will pass through. You can start this command by going to the **Home** tab, locating the **Modify** panel, then selecting the **Offset** button:

- You will see the following AutoCAD prompts:

```
Current settings: Erase source=No Layer=Source
OFFSETGAPTYPE=0
Specify offset distance or [Through/Erase/Layer] <Through>:
```

4.2.1 Offsetting Using the Offset Distance Option

- If you know the distance between the object and the new parallel copy, then input this value, select the original object, and then click on the side you want the new object to go on. You will see the following prompts:

```
Specify offset distance or [Through/Erase/Layer] <Through>:
Select object to offset or [Exit/Undo] <Exit>:
Specify point on side to offset or [Exit/Multiple/
Undo] <Exit>:
```

- This will allow you to create a single offset. To create more offsets using the same offset command, select another object, and follow the same steps again. Pressing [Enter] or right-clicking will end the command.

4.2.2 Offsetting Using the Through Option

- If you don't know the offset distance, but you know a point in the drawing the new parallel object will pass through, this option will help you accomplish your mission. You will see the following AutoCAD prompts:

```
Specify offset distance or [Through/Erase/Layer] <Through>:
Select object to offset or [Exit/Undo] <Exit>:
Specify through point or [Exit/Multiple/Undo] <Exit>:
```

- This will allow you to create a single offset. To create more offsets using the same offset command, select another object, and follow the same steps again. Pressing [Enter] or right-clicking will end the command.
- Here is an example:

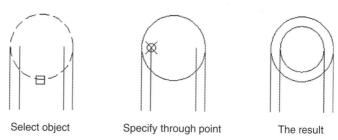

Select object   Specify through point   The result

4.2.3 Using the Multiple Option

- You can use the Multiple option to repeat the offset distance or the Through option in the same command repeatedly by clicking on the side of the offset, or by specifying a new through point. The prompts for the Multiple option are:

```
Specify through point or [Exit/Multiple/Undo] <Exit>: M
Specify point on side to offset or [Exit/Undo] <next
object>:
```

- While offsetting keep the following in mind:
  - If you make any mistake use the Undo option.
  - AutoCAD remembers the last offset distance used and will save it in the file.
  - When offsetting an arc or circle, the new arc and circle will share the same center point, and the result will be a smaller or larger arc or circle.

- If you offset a closed polyline, the output will be smaller or larger.
- Offset includes the Automatic Preview feature, which will show you the result before you accept it.

PRACTICE 4-1

Offsetting Objects

1. Start AutoCAD 2014.
2. Open **Practice 4-1.dwg**.
3. Offset the circle to inside by distance = 0.5.
4. Offset the rightmost vertical line to the inside using the Through option and the midpoint of the small horizontal line. Do the same for the left side. See the following illustration:

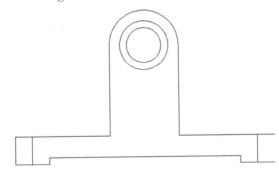

5. Using the two newly created lines offset each line to its right and its left by distance = 0.5.
6. You should have the following:

7. Save and close the file.

PRACTICE 4-2

Offsetting Objects

1. Start AutoCAD 2014.
2. Open **Practice 4-2.dwg**.
3. Offset the polyline that represents the outside edge of the outer wall by distance = 1'-0".
4. Explode the inner polyline.
5. Offset the yellow line representing one of the stair steps 10 times using distance = 1'-6" and the Multiple option.
6. Offset the vertical line at the right to the left using the Through option to pass through the inner right end point of the arc.
7. Offset the newly created line to the right by distance = 6".
8. You will get the following:

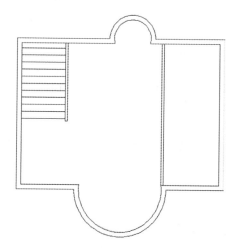

9. Save and close the file.

4.3 FILLETING OBJECTS

- The goal of the Fillet command is to create neat intersections. You should set the value of the radius as the first step. If it is 0 (zero) then you can only use Fillet between two lines and the command will extend/

trim the lines to the proposed intersection point. But if the value of the radius is greater than 0 (zero) then Fillet can use lines and circles to fillet these objects with an arc. See the following:

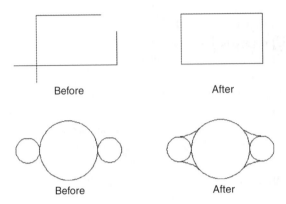

Before   After

Before   After

- While working with a radius greater than zero, you can select between **Trim**, which allows you to trim the original objects or **No trim**, which allows the original objects to stay as is. In both cases, you will be able to see the arc when hovering over the second object and you can make sure that it is the right value!
- Here is an example:

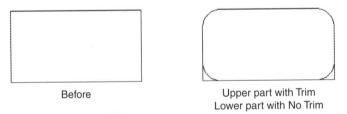

Before   Upper part with Trim
                         Lower part with No Trim

- To start this command, go to the **Home** tab, locate the **Modify** panel, then select the **Fillet** button:

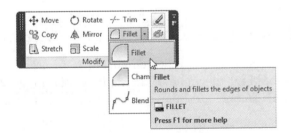

- You will see the following AutoCAD prompts:

```
Current settings: Mode = TRIM, Radius = 0.0000
Select first object or [Undo/Polyline/Radius/Trim/ Multiple]:
```

- You should always check the first line of this command because it will report the current value of Radius, and you can decide whether to keep it or change it. Type **r** or right-click and select the Radius option to set the new value of the radius. You will see the following prompt:

```
Specify fillet radius <0.0000>:
```

- Type **t** or right-click and select **Trim** to change the mode to **Trim** or **No trim**. You will see the following prompt:

```
Enter Trim mode option [Trim/No trim] <Trim>:
```

- The Fillet command allows only a single fillet per command. To make multiple fillets in the same command, simply change the value to **Multiple**. If you make a mistake use the **Undo** option to undo the last action. To end the command press [Enter].
- You can do two important things while using the Fillet command:
  - You can fillet two parallel lines regardless of the current radius value.
  - You can fillet any two lines with radius = 0 regardless of the current value of radius by holding the [Shift] key.
- If you use the Multiple option, you can use different radius values in the same command.

PRACTICE 4-3

 Filleting Objects

1. Start AutoCAD 2014.
2. Open **Practice 4-3.dwg**.

3. Using the Fillet command try to get the following final result:

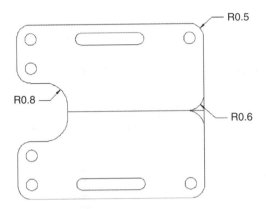

4. Save and close the file.

4.4 CHAMFERING OBJECTS

- The Chamfer command will create neat intersections as well, but with this command you should set the value of Distance (or Distance and Angle) as the first step. If it is 0 (zero) then you can use Chamfer to extend/trim the lines to the proposed intersection point. But if the value of Distance is greater than 0 (zero) then Chamfer will create a sloped edge between the two lines. While working with a distance greater than zero, you can select between **Trim**, which allows you to trim the original objects or **No trim**, which allows the original objects to stay as is. In both cases, you will be able to see the chamfer line when hovering over the second object and you can make sure that it's the right value.
- To create the sloped edge, use one of the following two methods:
  - Distance (two distances)
  - Distance and Angle

4.4.1 Chamfering Using the Distance Option

- You will see the following two prompts when you chamfer using the Distance option:

```
Specify first chamfer distance <0.0000>:
Specify second chamfer distance <0.0000>:
```

- There will be two different cases:

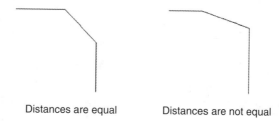

Distances are equal   Distances are not equal

4.4.2 Chamfering Using Distance and Angle Options

- You will see the following two prompts when you chamfer using Distance and Angle:

```
Specify chamfer length on the first line <0.0000>:
Specify chamfer angle from the first line <0>:
```

- Set the length (which will be cut from the first object selected) and an angle. See the following illustration:

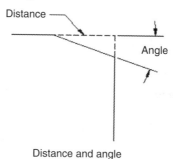

Distance and angle

- To start this command, go to the **Home** tab, locate the **Modify** panel, then select the **Chamfer** button:

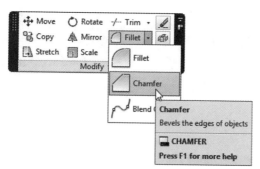

- You will see the following prompts:

  ```
  (TRIM mode) Current chamfer Dist1 = 0.0000, Dist2 = 0.0000
  Select first line or [Undo/Polyline/Distance/Angle /Trim/
  mEthod/Multiple]:
  ```

- You should always check the first line of this command because it will report the current value of the method used (whether Distance or Distance and Angle) and the current values. Accordingly, you can decide to keep it or change it.
- The other options, Multiple, Trim, and Undo, are identical to what we learned for the Fillet command. The Method option is used to select the default method to be used in the chamfering process.
- Another similarity to the Fillet command is you have to hold the [Shift] key, which allows you to chamfer by extending/trimming the two lines, regardless of the current distance values.

- Trim and Untrim in Chamfer will affect the Fillet command and vice-versa.

PRACTICE 4-4

Chamfering Objects

1. Start AutoCAD 2014.
2. Open **Practice 4-4.dwg**.
3. Use the Chamfer command to make the shape look like the following:

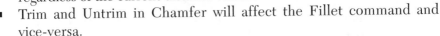

4. Save and close the file.

4.5 TRIMMING OBJECTS

- This command allows you to remove part of an object based on cutting edge(s). In order to successfully complete this command, you have to select the cutting edges first, press [Enter], and then select the part of the objects you want to remove.
- The following example illustrates the process of trimming:

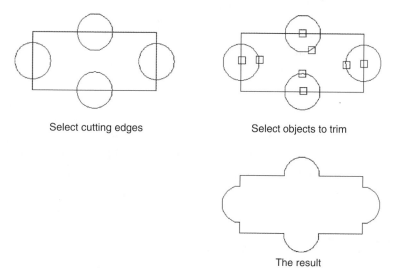

Select cutting edges   Select objects to trim

The result

- To issue this command, go to the **Home** tab, locate the **Modify** panel, then select the **Trim** button:

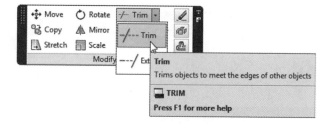

- You will see the following prompts:

```
Current settings: Projection=UCS, Edge=Extend
Select cutting edges ...
Select objects or <select all>:
```

- The first line gives you the current settings. The second line asks you to select the cutting edges, which is the first step of the Trim command. When you're done press [Enter] or right-click. You can also use the Select All option, which will select all the objects as the cutting edges. This can be done by pressing [Enter].
- You will see the following prompt either way:

```
Select object to trim or shift-select to extend or
[Fence/Crossing/Project/Edge/eRase/Undo]:
```

- You can now start clicking on the objects you want to remove. You have three ways to do that:
  - Select the objects one by one.
  - Click on an empty space and go to the left to start Crossing mode to select the objects to be removed collectively.
  - Type F to start Fence option (discussed before in Chapter 3).
- If you made any mistake simply type u to undo the last trim.
- While you are trimming you may get orphan objects, but you can get rid of them with the eRase option (type r).

PRACTICE 4-5

 Trimming Objects

1. Start AutoCAD 2014.
2. Open **Practice 4-5.dwg**.
3. Using the Trim command try to make the shapes look like the following:

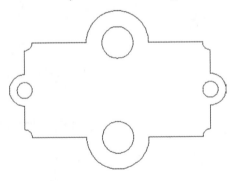

4. Save and close the file.

4.6 EXTENDING OBJECTS

- This command allows you to extend an end of an object to the boundary edge(s). In order to successfully complete this command, you have to select the boundary edges first, press [Enter], and then select the ends of the objects you want to extend.
- See the following illustration:

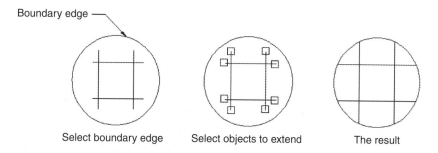

Select boundary edge   Select objects to extend   The result

- To issue this command, go to the **Home** tab, locate the **Modify** panel, then select the **Extend** button:

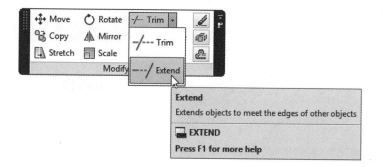

- You will see the following prompts:

```
Current settings: Projection=UCS, Edge=Extend
Select boundary edges ...
Select objects or <select all>:
```

- The first line gives you the current settings. The second line asks you to select the boundary edges, which is the first step of the Extend command; when done press [Enter] or right-click. You can also use the **select all** option, which will select all the objects to act as boundary edges. This can be done by pressing [Enter]. Either way, you will see the following prompt:

```
Select the object to extend or shift-select to trim
or[Fence/Crossing/Project/Edge/Undo]:
```

- Now you can start now clicking on the part of objects you want to extend. You can do this one of three ways:
  - Select the object ends one by one.
  - Click on an empty space and go to the left to start Crossing mode to select the objects to be removed collectively.
  - Type F to start the Fence option (discussed in Chapter 3).
- If you make a mistake simply type u to undo the last extend.
- The last feature in both Trim and Extend commands provides the ability to use each command while you are using the other. See the following example:

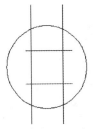   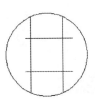

Select the circle as cutting edge   Trim lines as shown   Hold [Shift] to convert Trim command to Extend command and select the other lines

PRACTICE 4-6

   Extending Objects

1. Start AutoCAD 2014.
2. Open **Practice 4-6.dwg**.

3. Using the Extend command (and Trim if needed) correct the architectural plan so it looks like the following:

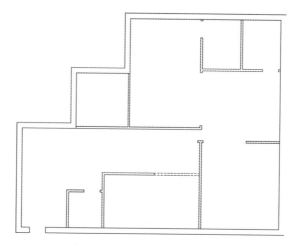

4. Save and close.

4.7 ARRAYING OBJECTS – RECTANGULAR ARRAY

- This command allows you to create replicates in a matrix fashion using rows and columns. The resulting shape will be one object, which can be edited.

4.7.1 The First Step

- To issue this command, go to the **Home** tab, select the **Modify** panel, and select the **Rectangular Array** button:

- The following prompt will be shown:

```
Select objects: 1 found
```

- This prompt asks you to select the desired objects. When you're done, press [Enter] and AutoCAD will immediately add a 3 row and 4 column grid with grips like the following:

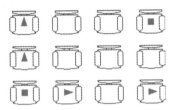

- Subsequently, you will see the following context tab called Array Creation:

- The following prompts will appear at the Command window:

```
Type = Rectangular Associative = Yes
Select grip to edit array or [ASsociative/Base point/
COUnt/Spacing/COLumns/Rows/Levels/eXit]<eXit>:
```

- AutoCAD asks you to select the grip to edit the array. You can do this two ways:
  - Using the Array Creation context tab
  - Using grips to set the number of rows, columns, distances between columns, distances between rows, and direction of arraying (downward or upward, right, or left)

4.7.2 Using the Array Creation Context Tab

- Using the Array Creation context tab is more convenient if you have specific measurements. You can do this as follows:
  - Use the Column panel to input two of three pieces of information: Columns (number of columns), Between (Distance between Columns), and Total (Total distance the columns will occupy). Make

sure you are consistent and use the same reference point to measure distances (from left to left, or from center to center). You also need to consider the arraying direction: positive distances mean upwards and negative distances mean downward.

- Use the Rows panel to input two of three pieces of information: Rows (number of rows), Between (Distance between Rows), and Total (Total distance the rows will occupy). Again, make sure to use the same reference point to measure distances (from top to top, or from center to center). Also take the arraying direction into consideration. Ignore Levels for 2D, because this is only for 3D.
- Associative means all objects resulting from the array will be considered a single object holding all the information used to build it.
- By default the first object will be considered the base point for the array, but AutoCAD allows you to select a different one.
- Click the Close Array button to end the command.

4.7.3 Editing a Rectangular Array Using Grips

- After you insert a rectangular array, click on any object to edit it. Grips will appear on the objects like the following:

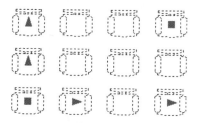

- Each one of these grips has a function:

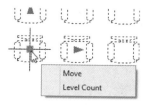

- This grip is for moving the whole array object, or to control the Level Count (3D only). Use [Ctrl] to browse between these options.

- This grip is for changing the column spacing.

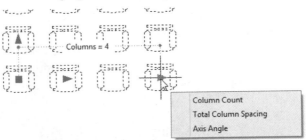

- This grip is for changing the Column Count, Total Column Spacing, or Axis Angle (to specify another angle other than horizontal and vertical). Use [Ctrl] to browse between these options.

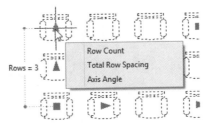

- This grip is for changing the Row Count, Row Spacing, or Axis Angle. Use [Ctrl] to browse between these options.

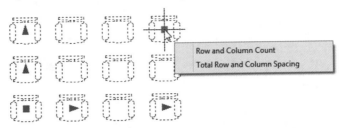

- This grip is for changing the Row and Column Count or the Total Row and Column Spacing. Use [Ctrl] to browse between these options.

4.7.4 Editing a Rectangular Array Using the Context Tab

- When you click a rectangular array a context tab called Array will appear, something like the following:

- This tab is almost identical to the Array Creation context tab, except for the Options panel, which includes three buttons:
  - Edit Source allows you to make changes to one of the objects arrayed, which will be reflected in all other arrayed objects.
  - Replace Item allows you to replace one or more of the arrayed shapes with another shape.
  - Reset Array will undo the effects of the Replace Items command.

4.7.5 Editing Rectangular Array Using Quick Properties

- When you click a rectangular array Quick Properties will be shown, something like the following:

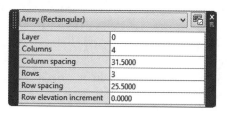

- You can change Columns number, Columns spacing, Rows number, and Rows spacing.

PRACTICE 4-7

Arraying Objects Using the Rectangular Array

1. Start AutoCAD 2014.
2. Open **Practice 4-7.dwg**.

3. Using the chair try to create a rectangular array like the following:

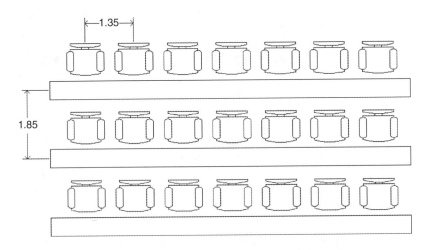

4. Exit the Array command.
5. Click the rectangular array, and using grips make the number of columns = 5 and distance between columns = 2.
6. Make layer Chair current.
7. Click the rectangular array to select it, click the Edit Source button, and select the upper-right chair (you can select any other one if you want). When the message appears, click OK, and you will notice that only the chair you selected is highlighted; zoom to it and add two diagonal lines.
8. A small panel at the right is displayed called Edit Array; click it and select the Save Changes button. Notice that all of the chairs reflect the changes.
9. Thaw layer Different.
10. Another chair will appear at the left.
11. Select the rectangular array.
12. Select the Replace Item button.
13. Select the new chair at the left, then press [Enter].
14. When AutoCAD asks about the base point for replacement objects, set the new base point using OSNAP (midpoint) and OTRACK to locate the center of the rectangle of the new chair.
15. Click all the chairs on the bottom row, then press [Enter] twice and then close the Array command.

16. You should have the following:

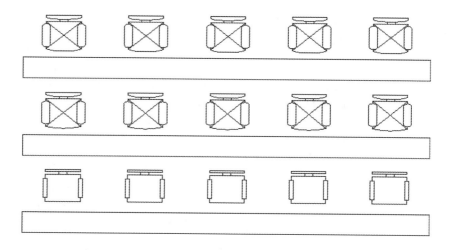

17. Save and close the file.

4.8 ARRAYING OBJECTS – PATH ARRAY

- This command allows you to create an array using an object such as a polyline, spline, arc, etc. To issue this command, go to the **Home** tab, locate the **Modify** panel, and select the **Path Array** button:

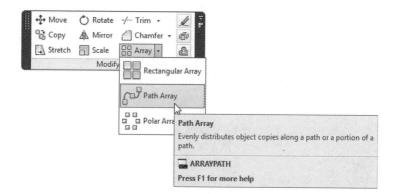

- The following prompts will be shown:

```
Select objects:
Type = Path Associative = Yes
Select path curve:
```

- The first prompt asks you to select the desired objects; once done, press [Enter]. The second prompt shows you that the type of array is path and that associativity is on. AutoCAD then asks you to select the path curve. Once selected, AutoCAD will show 10 objects arrayed using the path. The following prompt will appear:

```
Select grip to edit array or [ASsociative/Method/Base
point/Tangent direction/Items/Rows/Levels/Align
items/Z direction/eXit]<eXit>:
```

- The Array Creation context tab appears and looks like the following:

- When arraying, you should consider three things:
  - Base point
  - Aligning objects with the path
  - Measure or Divide
- Let's see how AutoCAD will handle the following case:

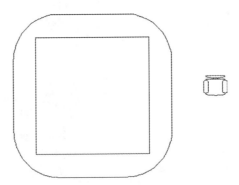

- We will select the chair as the object to be arrayed, and the outer polyline to be the path:

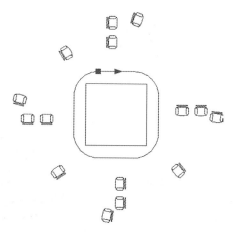

- Now we will go to the context tab and select the Base point option to specify a new base point (which will be the center of the chair):

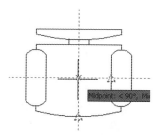

- The following is the result after specifying the new base point:

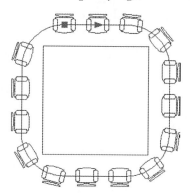

- Of course, the Align button is on by default, but if you turn it off, this is what you will get:

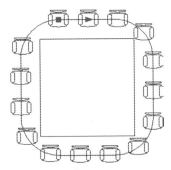

- If you select the Divide option, AutoCAD will divide the path equally for all the objects, but if you select Measure, you will specify the distance between each object. The above was produced using Measure, and the following by Divide:

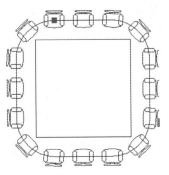

- After you specify the three factors, you will be able to set the Items (number of items), Between (the distance, which will be off in case of Divide), and Total (the total distance). You can also add rows, and you will get something like the following:

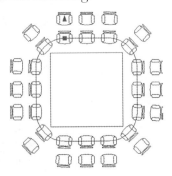

- After you create a path array, you can select the array for editing. You will see three grips (this will be true only for Divide; in Measure, you will see an extra arrow):

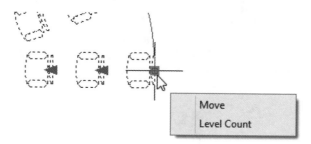

- You will use these to control the movement and the level count.
- The second one is:

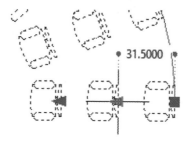

- This grip will specify distance between rows. The third one is:

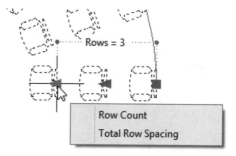

- It will control the Row Count, and the Total Row Spacing.

- A new context tab called Array will appear, which looks like the following:

- This is identical to what we discussed earlier.
- Also, when you select Path array, you will see something like the following Quick Properties:

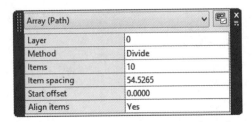

- In the Quick Properties, you can change the Method (Divide or Measure), the number of items, the item spacing, the Start offset (not to start from the first point of the path), and finally whether to align items or not while arraying them.

PRACTICE 4-8

Arraying Objects Using the Path Array

1. Start AutoCAD 2014.
2. Open **Practice 4-8.dwg**.
3. Using path array try to create an array, keeping the following in mind:
   a. Number of items = 22
   b. Base point = Center of the Chair (use OSNAP + OTRACK)
   c. Divide
   d. Align
   e. Number of rows = 2
   f. Distance between rows = 1.5
4. Erase the outer polyline.

5. You should get the following result:

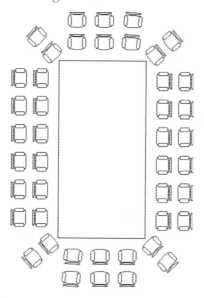

6. Using grips you can make the number of rows = 3 and 4 and back to 2.
7. Save and close the file.

4.9 ARRAYING OBJECTS – POLAR ARRAY

- This command allows you to duplicate objects in a circular fashion. To issue this command, go to the **Home** tab, locate the **Modify** panel, then select the **Polar Array** button:

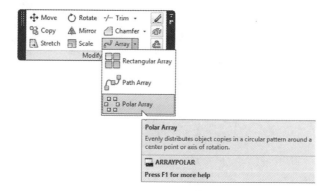

- The following prompts will be shown:

```
Select objects:
Type = Polar  Associative = Yes
Specify center point of array or [Base point/Axis of
rotation]:
```

- The first prompt asks you to select the desired objects; once done, press [Enter]. The second prompt shows that the type of array is polar and that associativity is on. The third line asks you to specify the center point of the array. Once you specify the center AutoCAD will create a polar array with 6 objects filling 360°. You will also see a new context tab called Array Creation, which looks like the following:

Home	Insert	Annotate	Layout	Parametric	View	Manage	Output	Plug-ins	Autodesk 360	Featured Apps	Express Tools	Array Creation	
	Items:	6		Rows:	1		Levels:	1					X
Polar	Between:	60		Between:	25.5000		Between:	1.0000	Associative	Base Point	Rotate Items	Direction	Close Array
	Fill:	360		Total:	25.5000		Total:	1.0000					
Type		Items			Rows ▾			Levels			Properties		Close

- Using this context tab you can specify the number of items, the angle between items, and the angle to fill. You can also specify the number of rows and the distance between rows. Using the Properties panel, choose whether this is going to be an associative array, specify a new base point for the object to be arrayed, specify whether to rotate items as you are copying them, and finally, the direction of arraying (CW or CCW).
- AutoCAD allows you to select the polar array for further editing. Once you select it, you will see something like the following:

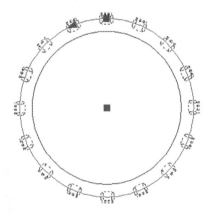

- The first grip will show something similar to the following:

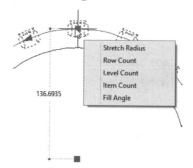

- You will see the current radius value along with a menu to help you: stretch the radius, change the row count, change the level count (for 3D only), change the item count, and finally change the fill angle. The following is an example of changing the row count:
- The second arrow grip will show the current angle between the first and second items.

- A new context tab called Array will appear, which looks like the following:

- This is identical to what we have already discussed.
- Also, when you click polar array you will see the following Quick Properties:

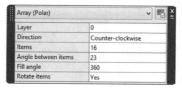

- In Quick Properties, you can change the direction of the array (CW or CCW), change the number of items, and angle between items, the total fill angle, and whether to rotate items while they are copied.

PRACTICE 4-9

Arraying Objects Using the Polar Array

1. Start AutoCAD 2014.
2. Open **Practice 4-9.dwg**.
3. Using polar array create the following shape:

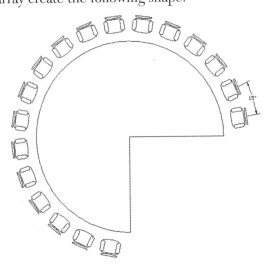

4. Save and close the file.

4.10 BREAK COMMAND

- This command will help you break any object into two objects by removing the portion between two specified points. To issue this command, go to the **Home** tab, locate the **Modify** panel, and select the **Break** button:

- You will see the following prompts:

```
Select object:
Specify second break point or [First point]:
```

- If selecting the object also specifies the first point, then select the second point. This will end the command. But if you think selecting is just selecting, and you didn't specify the first point, then type **F**, and you will see the following two prompts:

```
Specify first break point:
Specify second break point:
```

- Using these two prompts, specify two points on the object and the Break command will end.
- AutoCAD offers the same command using a different technique, breaking on the same point. To issue this command, go to the **Home** tab, locate the **Modify** panel, then select the **Break at Point** button:

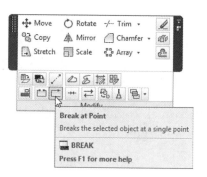

- You will see the following prompts:

```
Select object:
Specify second break point or [First point]: _f
Specify first break point:
Specify second break point: @
```

- After you select the desired object, it will flip automatically to allow you to specify the first point. At the second point prompt, AutoCAD responds with @, which means "use the same first point."

■   If the object you are breaking is a circle, make sure to specify the two
        points CCW. See this example:

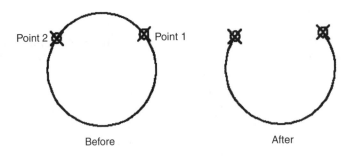

Before   After

PRACTICE 4-10

Breaking Objects

1. Start AutoCAD 2014.
2. Open **Practice 4-10.dwg**.
3. In order to locate the meeting table exactly at the center of the room
    horizontally and vertically, we need to break the upper horizontal line
    and the left vertical line.
4. Using the Break command, break the upper horizontal line from the
    points shown below:

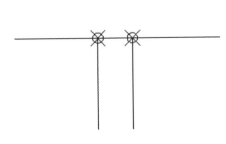

5. Using the Break command, break the left vertical line from the points shown below:

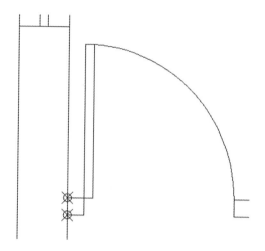

6. Using the Move command, move the meeting table to the center of the room using OSNAP and OTRACK

7. You will get the following:

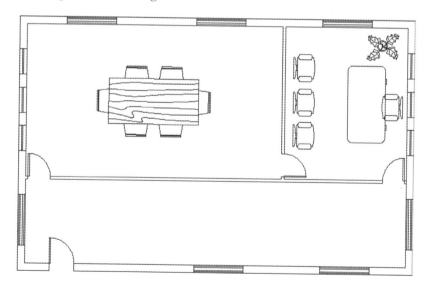

8. Save and close the file.

NOTES:

CHAPTER REVIEW

1. You can fillet using more than one fillet radius using the same command.
   a. True
   b. False
2. There are three types of arrays.
   a. True
   b. False
3. If you hold _____ while filleting, you will get radius = 0.0 regardless of the current radius.
4. Which one of the following is *not* related to the Chamfer command?
   a. Distance 1 and Distance 2
   b. Distance and Angle
   c. Distance and Radius
   d. Trim and No trim
5. You can offset using offset distance and _____.
6. In the Trim command, you will select the _____ as a first step and then you will press [Enter].
7. While using the Extend command if you hold _____ you convert the command to Trim:
   a. [Ctrl]
   b. [Ctrl] + [Shift]
   c. [Shift]
   d. [Alt] + [Ctrl]
8. While using the Break command you can break at the same point, using which character:
   a. $
   b. #
   c. @
   d. &

CHAPTER REVIEW ANSWERS

1. a
3. [Shift]
5. Through
7. c

Chapter 5

LAYERS AND INQUIRY COMMANDS

In This Chapter
◇ What are layers in AutoCAD?
◇ How to create and set layer properties
◇ What are layer controls?
◇ How to use Quick Properties and Properties
◇ Inquiry commands

5.1 USING LAYERS IN AUTOCAD

- Layers are the most important way to organize and control your Auto-CAD drawings. Managing layers means managing the drawing. So, what are layers in AutoCAD? Layers are much like a transparent piece of paper in which you draw part of the drawing using a certain color, line-type, and lineweight. Each object on a layer holds the properties of the layer; that is, the object will have the same color, linetype, and lineweight of the layer it resides in. This setting is called **BYLAYER**, which means we control the drawing through the layers rather than the objects.
- Each layer should have a name, which is the first step of the creation process. Layer names:
  - Should not exceed 255 characters
  - Can include all letters (upper or lowercase)
  - Can include any number
  - Can include the (-) hyphen, (_) underscore, and ($) dollar sign
- A unique layer exists in all AutoCAD drawings called 0 (zero). This specific layer can't be deleted or renamed. Other layers can be deleted and renamed.

- The layer at the top of the pile is the only layer we can draw on; this layer is called the current layer. So, as a rule of thumb, make the desired layer current first and then start drawing. To start building up your layers, go to the **Home** tab, locate the **Layers** panel and then click the **Layer Properties** button:

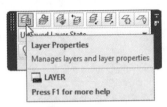

- You will see the following palette, called **Layer Properties Manager**:

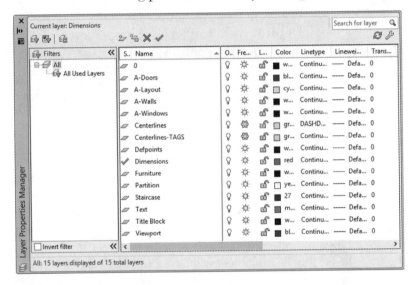

- Palettes in AutoCAD have more features than a normal dialog box:
  - The dialog box has two states: displayed on the screen or closed. The palette can also be displayed but hidden, which keeps you from having to open and close the box:

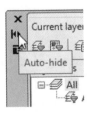

- Contrary to most dialog boxes, palettes can be resized horizontally, vertically, and diagonally:

- Dialog boxes can't be docked, but palettes can be docked on the four sides of the screen.

5.2 CREATING AND SETTING LAYER PROPERTIES

- In this section, we will learn how to:
  - Create a new layer
  - Set a color for a layer(s)
  - Set the linetype for a layer(s)
  - Set a lineweight for a layer(s)
  - Set the current

5.2.1 How to Create a New Layer

- This command allows you to add a new layer to the current drawing. Using the **Layer Properties Manager**, click the **New Layer** button:

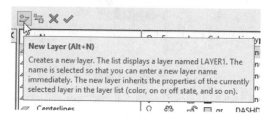

- This will create a new layer with the temporary name *Layer1*. The **Name** field will be highlighted. Enter the desired name of the layer. You should always stick to a good naming convention; that is, a layer containing doors should be named Doors.
- **NOTE** ▪ All the settings discussed in the following require you to select layer(s) in the **Layer Properties Manager**. You select layers in AutoCAD just like any other software running under the Windows OS. You can hold the [Ctrl] key and/or [Shift] to select multiple layers.

5.2.2 How to Set a Color for a Layer(s)

- You can use one of the 256 colors available in AutoCAD. The first seven colors can be set using the color's name or number:
  - Red (1)
  - Yellow (2)
  - Green (3)
  - Cyan (4)
  - Blue (5)
  - Magenta (6)
  - Black/White (7)
- Other colors should be set using only the number.
- To set the color for a layer, take the following steps:
  - Using the **Layer Properties Manager**, select the desired layer(s).
  - Using the **Color** field, click the icon of the color, and you will see the following dialog box:

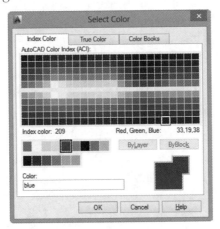

  - Select the desired color (or you can type the name/number in the **Color** field), then click the **OK** button to end this action.
- Another way to set up (or modify) a layer's color is by using the pop-up list in the **Layers** panel, as shown below:

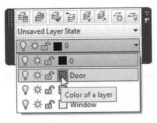

5.2.3 How to Set the Linetype for a Layer(s)

- There are two linetype files that come with AutoCAD 2014: *acad.lin* and *acadiso.lin*. Since these two files are not adequate for all types of engineering, designing, and drafting, you should consider buying more linetype files available on the Internet.
- Contrary to colors, linetypes are not loaded to the current file. So you should load the desired linetypes as needed. To set the linetype for a layer, take the following steps:
  - Using the **Layer Properties Manager**, select the desired layer(s).
  - Using the **Linetype** field, click the name of the linetype, and you will see the following dialog box:

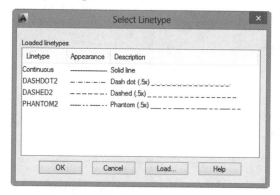

  - If your desired linetype is listed, select it. If it is not, you need to load it. Click the **Load** button, and you will see the following dialog box:

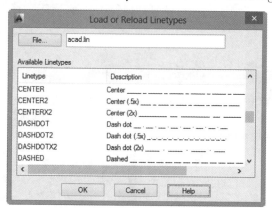

  - Browse for your desired linetype, select it, then click **OK**. Now the linetype is loaded, and you can select it and click **OK**.

5.2.4 How to Set a Lineweight for a Layer(s)

- This option allows you to set the lineweight for a layer(s). All objects in AutoCAD (except polylines with width) have a default lineweight, which is 0 (zero), but you can set the lineweight for objects through their layers. To do so take the following steps:
  - Using the **Layer Properties Manager**, select the desired layer(s).
  - Under the field **Lineweight**, click the lineweight icon, and the following dialog box will appear:

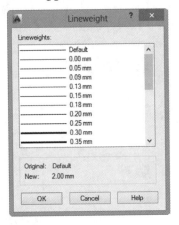

  - Select the desired Lineweight, and click **OK**.
  - To see the lineweight use the status bar and click the **Show/Hide Lineweight** button on:

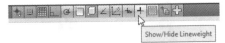

5.2.5 How to Set the Current Layer

- There are several ways to make a layer the current layer.
- The easiest way is to use the layer pop-up list in the Layers panel, as shown here:

- Another way is to use the **Layer Properties Manager** palette, then double-clicking the status of the desired layer's name.
- The longest way is to use the **Layer Properties Manager** palette; select the desired layer and then click the **Set Current** button:

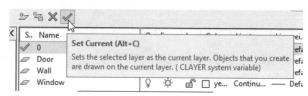

PRACTICE 5-1

   Creating and Setting Layer Properties

1. Start AutoCAD 2014.
2. Open **Practice 5-1.dwg**.
3. Create a new layer and call it Centerlines; the color should be yellow and the linetype should be Center. Make it current and draw two lines, from 8,14 to 16,14, and the other line from 12,11 to 12,17.
4. Create another layer and call it Hidden; the color should be 9 and the linetype should be Dashed. Make it current and draw two lines, from 3.75,4.5 to 3.75,8.5, and the other line from 20.25,4.5 to 20.25,8.5.
5. You should have the following shape:

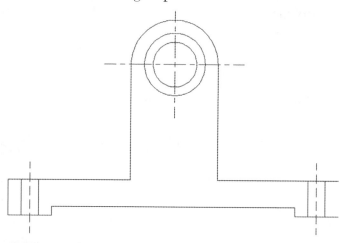

6. Save and close the file.

5.3 LAYER CONTROLS

- The commands discussed here help you have full control over your layers since they are the main way you control your drawings. We will learn how to control the visibility of layers, lock layers, plot layers, delete layers, rename layers, etc.

5.3.1 Controlling Layer Visibility, Locking, and Plotting

- AutoCAD provides controls to show/hide layers (Freeze and Off), lock and unlock layers, plot and not plot layers. See the following:
- The first example is layer "Title Block":

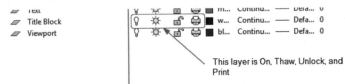

- The second example is layer "Staircase":

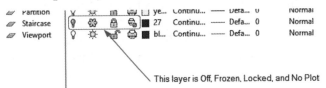

- When a new layer is created, by default, it will be On, Thawed, Unlocked, and Plotted. You can hide the contents of layers by turning them off or freezing them. Freeze has a deeper effect than Off, since objects in a frozen layer will not be considered in the drawing, and the drawing size will be temporarily smaller (you can use Freeze to make the drawing size smaller if the drawing loads slowly).
- If you tried to freeze the current layer, AutoCAD will show the following message:

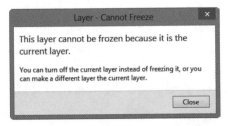

- But if you try to turn off the current layer, you will see the following message:

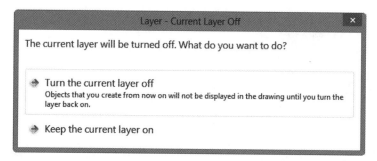

- If you lock a layer, then objects reside in it, and it will not be selected with any modifying commands. Objects in a locked layer will be faded, and when you get close to them you will see a small lock icon appear that tells you this layer is locked. See the following illustration:

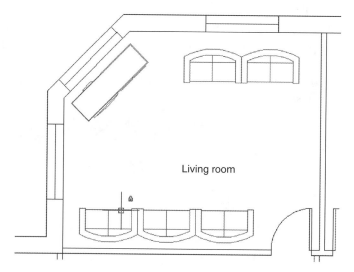

- If you choose No Plot for a layer, then objects in this layer will be displayed but not plotted.
- On/Off, Thaw/Freeze, Lock/Unlock can be controlled using the pop-up list in the Layer panel and Layer Properties Manager, while Plot/No Plot can be controlled only in the Layer Properties Manager.

5.3.2 Deleting and Renaming Layers

- AutoCAD will not delete a layer that contains objects. AutoCAD will delete only empty layers. In order to delete layers, take the following steps:
  - In the **Layer Properties Manager** palette, select the desired layer(s).
  - Press [Del] on the keyboard, or click on the **Delete Layer** button:

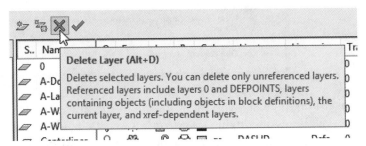

- In the **Layer Properties Manager,** you can rename layers. Take the following steps:
  - In the **Layer Properties Manager** palette, select the desired layer.
  - Click the layer name once, and the name will be highlighted for editing. Type new name, then press [Enter].
- If you tried to delete a layer containing objects, you will see the following message:

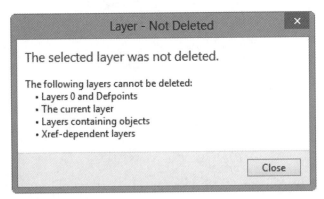

- To rename a layer, take the following steps:
  - Select the desired layer.

- Click the name a single time, and you will see the following:

- The name will become editable; type in a new name, and press [Enter].

5.3.3 How to Make an Object's Layer the Current Layer

■ This is the fastest way to make a layer the current layer! After issuing the command, select an object that resides in this layer, and follow the steps below:

- Go to the **Home** tab, locate the **Layers** panel, and then click the **Make Object's Layer Current** button:

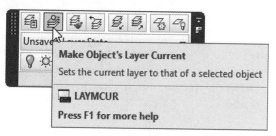

- You will see the following prompt:

```
Select object whose layer will become current:
```

- Select the desired object. See what the current layer is. You will find it became the object's layer (even without you knowing the object's layer).

5.3.4 How to Undo Layers Actions Only

■ The function to undo the layers actions only is called **Layer Previous**. This function will help you restore previous states of the layers (such as Freeze, Thaw, On, Off, etc.) without affecting other parts of the drawing or other modifying commands. Take the following steps:

- Change the layer states as needed.

- Go to the **Home** tab, locate the **Layers** panel, then click the **Previous** button:

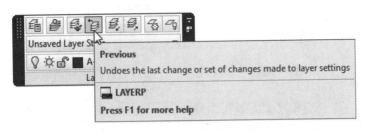

■ In the Command window, you will see the following message:

```
Restored previous layer status
```

5.3.5 Moving Objects from One Layer to Another

■ All similar objects must reside in the same layer, but mistakes occur happen. If you draw on the wrong layer, and you want to move the objects to the right layer, you can use the **Match** command. Do the following:
- Go to the **Home** tab, locate the **Layers** panel, then click the **Match** button:

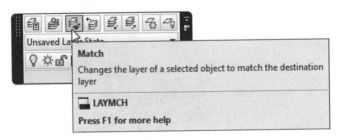

■ The following prompts will appear:

```
Select objects to be changed:
Select object on destination layer or [Name]:
```

■ This command contains two prompts; using the first you will select the object mistakenly drawn in the wrong layer, then at the second prompt, you will either select the object that resides in the right layer, or simply type its name.

- When you're done you will see something like the following in the Command window:

```
8 objects changed to layer "Dimensions"
```

5.4 USING THE LAYER PROPERTIES MANAGER

- While you are in the **Layer Properties Manager** palette you can do several things. If you select a layer and then right-click, you will see a menu like the following:

- In this menu, you can do any or all of the following:
  - Set the current layer
  - Create a new layer
  - Delete a layer
  - Select All layers
  - Clear the selection
  - Select All but Current
  - Invert the Selection (unselect the selected, and vice-versa)

■ You also have the option to show or hide the filter tree:
  • Show Filter Tree (by default it is turned on)
  • Show Filters in Layer List (by default it is turned off)
■ If you turned off **Show Filter Tree**, you will be allowed to see more information about your layers. See the following:

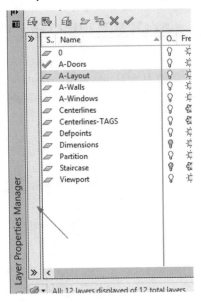

■ Another way to do the same thing is by clicking the small arrows at the top-right part of the filter tree pane:

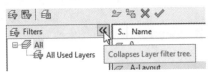

PRACTICE 5-2

Layer Controls

1. Start AutoCAD 2014.
2. Open **Practice 5-2.dwg**.
3. Make layer 0 the current layer.
4. Freeze layers Centerlines and Hatch.

5. Lock layers: Furniture. What happened to the color? _____
6. Get closer to any object in the Furniture layer, and notice the icon that appears when you get closer.
7. Try to erase one of the objects in layer Furniture. What message do you get? _____
8. Using the Layer Properties Manager, try to freeze layer 0 (the current layer). What message do you get? _____

   _____
9. Try to rename layer 0. What message do you get?_____

   _____
10. Rename layer Partition as Inside Wall.
11. Using the Match command, select the two doors at the bottom (their colors are black and white) to match one of the blue doors.
12. Using the Layer Properties Manager, select all the layers, then change their color to black or white), then close the Layer Properties Manager.
13. Click Layer Previous. What happened? _____
14. Click the Make Object's Layer Current button, and click one of the yellow lines. What is the current layer now? _____
15. Save and close the file.

5.5 CHANGING AN OBJECT'S LAYER, QUICK PROPERTIES, AND PROPERTIES

- In this section, we will learn to control object's properties.

5.5.1 Reading Instant Information About an Object

- AutoCAD provides you instant information about any object when your hover over the object. See the following illustration:

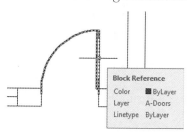

- You can see that AutoCAD is showing the type of object (Line), its color, the layer it's in, and its linetype.

5.5.2 How to Move an Object From a Layer to Another Layer

- All objects in AutoCAD should reside in a layer. To move an object from one layer to another, take the following steps:
  - Click the desired object(s).
  - Go to the **Home** tab, locate the **Layers** panel, and find the layer name in the pop-up list. Sometimes this will be blank; this happens when your desired objects reside in different layers. Click the layer's pop-up list and select the new layer name.
  - Press [Esc] once to deselect all the selected objects.

5.5.3 What is Quick Properties?

- This is the first of two commands that allow you to change the properties of selected objects. An object's properties differ depending on the object type, and the line properties are different depending on if it's an arc, circle, or polyline.
- **Quick Properties** pop up on the screen whenever you select objects without issuing a command (using grips). You can turn it off using the status bar, as shown below:

- When you select object(s) you will see the following (circle as example):

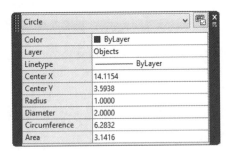

- You will see information including:
  - Color
  - Layer
  - Linetype

- Coordinates of Center point (X & Y)
- Radius and Diameter
- Circumference and Area

■ If you select more than one object of the same type, you will see this:

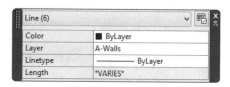

■ But if you select more than one object of different types, you will see this:

■ You will see All (number), which means you are seeing eight objects (in our example) from different types, but you can see a breakdown of the selected objects by clicking the pop-up list, as shown below:

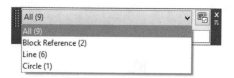

■ When you select a single object type, you can change the general and specific object's properties.
■ You should note that while you are at Quick Properties, whenever you select a different layer, color, linetype, etc., you will see the effect of your change concurrently. This also applies to the next command.

5.5.4 What is Properties?

■ As the name indicates, Quick Properties is a fast way to change the properties of an object(s). But the **Properties** command is much more comprehensive; all the rules we discussed concerning Quick Properties are applicable here as well.
■ To issue **Properties** command, select the desired object(s), right-click, and then select **Properties**.

- You can also double-click any object to get the Properties palette. Accessing Properties this way has two drawbacks; first, it will work for a single object only, and second, some objects such as polylines, blocks, hatches, and text will interpret this action as an edit. Either way, you will see something like the following:

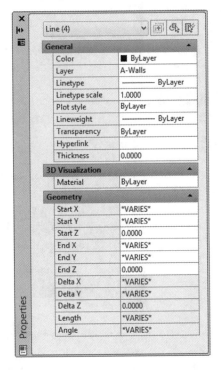

- As you can see, there is much more information displayed here than in Quick Properties, which gives you the chance to make more changes to your selected object(s).

- Some of the information is shaded, which means you cannot change it.

PRACTICE 5-3

Changing an Object's Layer, Quick Properties, and Properties

1. Start AutoCAD 2014.
2. Open **Practice 5-3.dwg**.

3. Hover over one of the circles of the centerlines. What is the name of the layer? _____

4. Select all the circles and text inside them and move them to layer Centerlines.

5. Change the properties of the circles to Continuous.

6. Delete layer Centerlines-TAGS.

7. Zoom to the lower door; you will find two red lines at the right and at the left. Move them from layer Dimensions to layer A-Walls.

8. Save and close the file.

5.6 INQUIRY COMMANDS – INTRODUCTION

- The main purpose of this set of commands is to allow you to measure the length between two points, find the radius of a circle or arc, and measure the angle, area, or volume of 3D objects.
- You will use this set of commands to make sure your drawing is correct and according to the design's intent.
- To issue these functions go to the **Home** tab and locate the **Utilities** panel.

5.7 MEASURING DISTANCE

- This command allows you to measure the distance between two selected points. To issue this command, go to the **Home** tab, locate the **Utilities** panel, and click the **Distance** button:

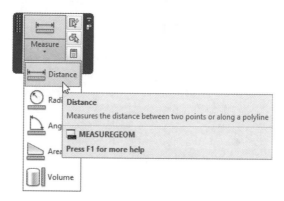

- You will see the following prompts:

```
Specify first point:
Specify second point:
```

- Select the desired points, and AutoCAD will display something like the following:

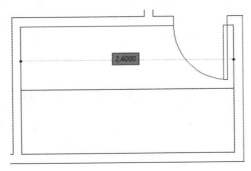

- You will see the following at the Command prompt:

```
Distance = 2.4, Angle in XY Plane = 0, Angle from XY
Plane = 0
Delta X = 2.4, Delta Y = 0.0000, Delta Z = 0.0000
```

5.8 CHECKING THE RADIUS

- This command allows you to check the radius (and parameter as well) of a drawn circle or arc. To issue this command, go to the **Home** tab, locate the **Utilities** panel, and select the **Radius** button:

- AutoCAD will display the following prompt:

```
Select arc or circle:
```

- Select the desired arc or circle, and you will see the following on the screen:

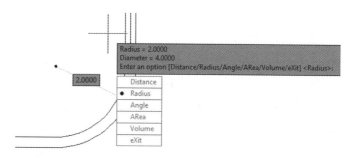

- You will also see the following in the Command window:

```
Radius = 2.000
Diameter = 4.000
```

5.9 MEASURING ANGLE

- This command allows you to measure the angle (between two lines, the included angle of an arc, or two points and the center of the circle). To issue this command, go to the **Home** tab, locate the **Utilities** panel, and then select the **Angle** button:

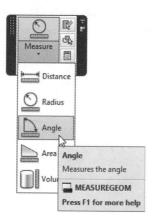

- You will also see the following prompt:

  ```
  Select arc, circle, line, or <Specify vertex>:
  ```

- Select the desired objects (whether two lines, an arc, or points on a circle), and AutoCAD will display the following:

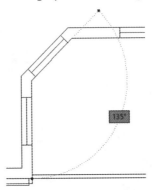

- You will also see something like the following in the Command window:

  ```
  Angle = 135°
  ```

5.10 MEASURING AREAS

- This command allows you to measure areas, whether simple (areas that have no islands inside) or complex areas (areas that have islands inside). AutoCAD can measure areas between points (assuming lines and arcs connecting them) or objects (such as circles, closed polylines, etc.). To issue this command, go to the **Home** tab, locate the **Utilities** panel, and then select the **Area** button:

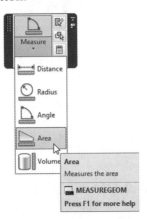

- You will also see the following prompt:

```
Specify first corner point or [Object/Add area/ Subtract
area/eXit] <Object>:
```

5.10.1 How to Calculate a Simple Area

- The definition of a simple area is any closed area without any objects (islands) inside it. AutoCAD will assume you want to measure a simple area if you start by specifying points or selecting objects. If you start specifying points, AutoCAD will assume there are either lines or arcs connecting them. For lines, you will see the following prompts:

```
Specify next point or [Arc/Length/Undo]:
Specify next point or [Arc/Length/Undo]:
Specify next point or [Arc/Length/Undo/Total] <Total>:
Specify next point or [Arc/Length/Undo/Total] <Total>:
```

- If you see an arc in your area, simply change the mode to Arc, and you will see prompts identical to the Polyline command. Keep specifying points of lines or arcs until you press [Enter]. You will see the Total value of the measured area in the Command window and the following:

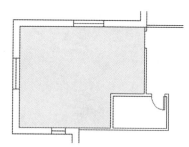

- You will see the following displayed in the Command window:

```
Area = 18.8366, Perimeter = 17.0416
```

- If you have a simple area and the parameter is a single object, you can select the object rather than specifying points. Either right-click and select the **Object** option, or type **o** in the Command window. You will see the following prompt:

```
Select objects:
```

- Select the object you want to measure, then press [Enter].

5.10.2 How to Calculate a Complex Area

- The definition of a complex area is any closed area with objects (islands) inside it. To tell AutoCAD you want to calculate a complex area, you <u>have</u> to start with either **Add area** or **Subtract area**.
- If you start with Add area or Subtract area, AutoCAD will start with area = 0, which allows you to add areas, then subtract areas as needed. For Add area mode you will see the following prompt:

```
Specify first corner point or [Object/Subtract area/eXit]:
```

- Specify the area using the same methods discussed above (points or object), then switch to Subtract area mode, and you will see the following prompt:

```
Specify first corner point or [Object/Add area/eXit]:
```

- AutoCAD will give you a sub-total after adding or subtracting each area. When you are done, press [Enter] twice to end the command, and you will get the final net area.
- You will see something like the following:

PRACTICE 5-4

Inquiry Commands

1. Start AutoCAD 2014.
2. Open **Practice 5-4.dwg**.
3. Freeze all layers except A-Walls (make layer A-Walls current, then select all layers except current and freeze them).
4. Measure the length of the slanted wall from inside, and enter the information given by AutoCAD here:
   a. Length = _____ (1.6971)
   b. Angle in XY plane = _____ (45 or 135)
   c. Delta X = _____ (1.2000)
   d. Dleta Y = _____ (1.2000)
5. Thaw layer Partition.
6. Measure the horizontal and vertical lengths of the room at the upper right part (you can use the Nearest and Perpendicular OSNAPs), and enter the measurements here:
   a. Horizontal Distance = _____ (5.05)
   b. Vertical Distance = _____ (3.9)
7. Measure the inside area of the room at the lower-right part of the plan, and enter the measurements here:
   a. Area = _____ (21.8016)
   b. Parameter = _____ (18.2416)
8. Save and close the file.

PRACTICE 5-5

Measuring Area

1. Start AutoCAD 2014.
2. Open **Practice 5-5.dwg**.
3. Calculate the net area of the shape without all the inside objects, and enter it here: Area = _____ (33.3426).
4. Save and close the file.

NOTES:

CHAPTER REVIEW

1. Which one of the following is *not* true about layer names?
   a. They should not exceed 256 characters
   b. Spaces are allowed
   c. $ is allowed
   d. $ is not allowed
2. You can't _____ the current layer.
3. You can _____ the current layer.
4. You can undo layer actions only.
   a. True
   b. False
5. The Area command can calculate only areas an area without any islands inside it.
   a. True
   b. False
6. The following are facts about layer 0 (zero) *except*:
   a. You can't it rename it.
   b. You can't set a new color for it.
   c. You can't delete it.
   d. It is in all AutoCAD files.
7. Delta X is one of the information properties in the_____ command.
8. Most objects will respond to _____ to display the Properties palette.

CHAPTER REVIEW ANSWERS

1. d
3. Turn off
5. b
7. Measure distance

6 BLOCKS AND HATCHES

Chapter

In This Chapter

◇ What are blocks and how do you define them?
◇ How to use (insert) blocks
◇ How to explode and convert blocks
◇ How to hatch in AutoCAD
◇ How to control hatches in AutoCAD
◇ How to edit hatches

6.1 WHAT ARE BLOCKS?

- In your daily work there will often be a shape(s) you need repeatedly, and there are two ways to get it. You can draw it each time you need it, or you can draw it once, and save it as a block (block will be a single object) and you can use (insert) it as many times as you wish in the current file or other files.
- There are many benefits to using blocks, including:
- Smaller file size because each block will be counted as a single object.
    - Standardization for the same company.
    - Speed of completing a drawing.
    - Using of Design Center and Tool Palettes.

6.2 HOW TO CREATE A BLOCK

- Use the following steps to create a block:
    - Draft the shape that you want to create a block from in layer 0 (zero). Layer 0 will enable the block to inherit the properties (color, linetype, lineweight) of the layer it resides in.

- Make sure to set the "Block unit," which allows AutoCAD to automatically scale the block to appear as the correct size in any other drawing.
- Draft the shape you want to create a block from in its real-world dimensions.

■ Let's assume we draw the following shape:

■ Now you are ready to issue the command. Go to the **Insert** tab, locate the **Block Definition** panel, and select the **Create Block** button:

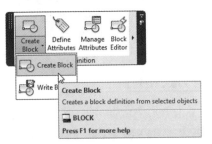

■ You will see the following dialog box:

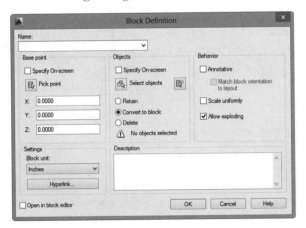

- Do the following:
  - Type a name for the block; it should not exceed 255 characters (use only numbers, letters, -, _, $, and spaces).
  - Specify the **Base point**, either by typing X, Y, Z coordinates, or clicking the **Pick point** button to input the base point graphically.
  - Click the **Select objects** button to select the desired objects.
  - From the drawn object, AutoCAD will create the needed block, but what do you do with the object afterwards? You should select one of the three choices: Retain (leave) the objects as they are, Convert them to a block, or Delete them:

- Select whether the block will be Annotative (a feature we will discuss in Chapter 9), whether to scale uniformly in both X and Y, and whether to allow the block to be exploded:

  - Select the **Block unit**. This will tell AutoCAD what each AutoCAD unit used in this block will equal. This will be helpful for the Automatic Scaling feature:

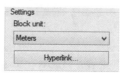

- Enter a block description.
- Click the checkbox "Open in a block editor" off because this is an advanced feature used for creating dynamic blocks.
  - Once you are finished inputting all the above data, click **OK**.
- Later, you will use (insert) the block, but this will only be a copy of the block. The original block definition will stay intact.

6.3 HOW TO USE (INSERT) BLOCKS

- After creating the block, you are ready to use (insert) the block in the current drawing. Make sure you are in the right layer, and the drawing is ready to accept the block (make sure the door openings are created before inserting the door, for example).
- Now you are ready to issue the command. Go to the **Insert** tab, locate the **Block** panel, and then click the **Insert** button:

- You will see the following dialog box:

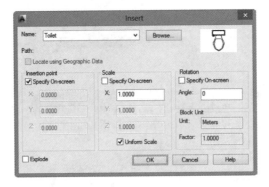

- If the block is created in the current file, click the pop-up list to select it. You can also click the **Browse** button to select a file and insert it in the current file as a block.

- You should specify the **Insertion point**, the **Scale**, and the **Rotation** angle either by using the **Specify On-screen** checkbox, or by entering the needed value.
- While using **Scale** you can insert mirror images of the block by using negative values. See the following illustration:

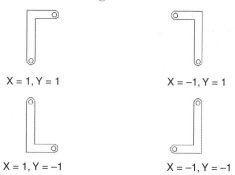

X = 1, Y = 1   X = −1, Y = 1

X = 1, Y = −1   X = −1, Y = −1

- While you are using Rotation angle, remember CCW is always positive.

6.3.1 Block Insertion Point OSNAP

- After you insert a block, click it to see the following:

- The whole block is one unit, and only one grip is highlighted which is the insertion point. There is a specific OSNAP to snap to this point called **Insertion** (or insert depending where you are looking). See the following illustration:

PRACTICE 6-1

Creating and Inserting Blocks

1. Start AutoCAD 2014.
2. Open **Practice 6-1.dwg**.
3. There are three shapes at the left drawn in layer 0. From these shapes, create three blocks, and name them Window, Single Door, and Double Door, making the block unit for the three blocks Meter.
4. Use the Insert command to insert the three blocks in the proper places using the proper layers, just like the following:

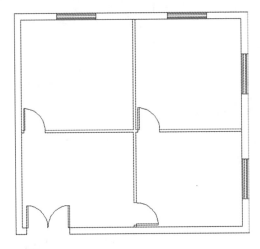

5. Save and close the file.

6.4 EXPLODING BLOCKS AND CONVERTING THEM TO FILES

6.4.1 Exploding Blocks

- You can explode blocks the same way you can explode polylines to lines and arcs. The Explode command will bring them back to their original objects. This practice is not recommended, however, since it's better to keep blocks as one object.

- To issue the command, go to the **Home** tab, locate the **Modify** panel, and then select the **Explode** button:

- You will see the following prompt:

```
Select objects:
```

- Select the desired block(s) and press [Enter] to end the command.

6.4.2 Converting Blocks to Files

- In earlier versions of AutoCAD, we used to convert all of our blocks to files in order to use them in other files. This practice was eclipsed by Design Center and Tool Palettes in more recent versions.
- To issue the command, go to the **Insert** tab and locate the **Block Definition**, then click the **Write Block** button:

- You will see the following dialog box:

- Select the **Block** option under **Source**. Select the name of the block. Under **Destination** input the file name and path and specify the insert unit. You can use the same dialog box to create a file from the entire drawing or from some of the objects in the current drawing.

PRACTICE 6-2

Exploding, Converting

1. Start AutoCAD 2014.
2. Open **Practice 6-2.dwg**.
3. Start the Block command, and select the Single Door block to convert it to a file, and save it in your practice folder.
4. Explode one of the single door insertions. Using Quick Properties, select one of the objects that resulted from the exploding process. In which layer does it reside? _____ Why? _____
   _____
5. Save and close the file.

6.5 HATCHING IN AUTOCAD

- AutoCAD can hatch closed areas and open areas (with the maximum distance defined by you).
- There are two hatch pattern files that come with AutoCAD: *acad.pat* and *acadiso.pat* (as you can see the hatch pattern file's extension is *.PAT).
- There are four pattern types:
  - Solid (single pattern covers the area with a single solid color)
  - Gradient (two gradient colors mixed together in several ways)
  - Pattern (several pre-defined patterns)
  - User-defined (the simplest pattern, parallel lines)

6.6 HATCH COMMAND: FIRST STEP

- This command allows you to put hatches in the drawing and control all of its properties. The preview is instant. To issue the **Hatch** command, go to the **Home** tab, locate the **Draw** panel, and then click the **Hatch** button:

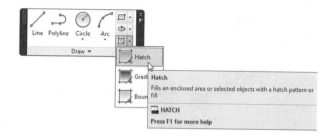

- You will see a new context tab called **Hatch Creation**. You will also see several panels (which will be discussed). Locate the **Properties** panel, at the top-left, and select the **Hatch Type** as shown below:

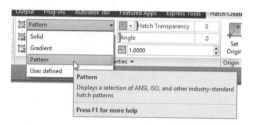

- Once you select the Hatch Type, locate the **Pattern** panel, and Auto-CAD will take you to the first pattern in the selected type. For instance, if you select **Gradient** for the Hatch Type, the first pattern in the Pattern panel will be GR_LINEAR, which is the first pattern in the gradient patterns:

- Now, simply go (without clicking) to the desired area to be hatched, and you will see the area filled. You now have two choices:
  - If you like the result, click to choose the area, then go to the **Close** panel, and click **Close Hatch Creation**.
  - If you don't like the result, you can change the properties of the hatch for a different result.

6.7 CONTROLLING HATCH PROPERTIES

- ■ If you clicked inside the area and you don't like the result, you need to alter the properties of the hatch. All of these properties exist in the **Properties** panel:
  - • **Hatch Color**: Specify the color of the hatch, or leave it as "Use Current":

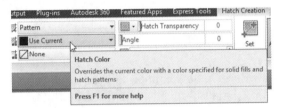

  - • **Background Color**: Specify the color of the background, or leave it as "None."

  - • **Transparency**: By default, all colors will be normal, but you can increase the value of transparency (maximum 90) to decrease the intensity of the color (hatch color and background color):

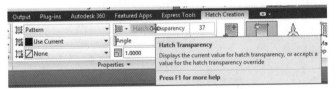

  - • **Angle**: Specify the angle of the hatch pattern (there will be no effect with Solid hatching):

- **Scale** (if you chose a user-defined type, this will be called Spacing); input the scale or spacing for the selected hatch pattern:

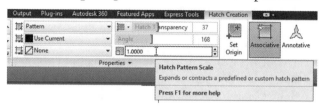

- **Hatch Layer Override**: By default, the hatch will reside in the current layer. Using this function you can specify the layer you want the hatch to reside in, regardless of the current layer:

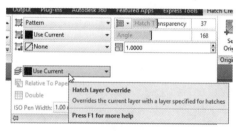

- **Double**: This option is only valid if the hatch type is user-defined. It controls whether the lines go in one direction or are crosshatched:

PRACTICE 6-3

Hatching and Controlling Hatch Properties

1. Start AutoCAD 2014.
2. Open **Practice 6-3.dwg**.

3. Make layer Hatch the current layer.
4. Start the Hatch command.
5. Make sure the following settings are correct:
    a. At the Properties panel, the Hatch Type is Pattern.
    b. At the Pattern panel, ANSI31.
6. Hover over any part of the drawing, and you will see a preview of the hatch. You will notice that the Scale is a little small, so increase it to 2.
7. Select Hatch background = Yellow.
8. Then select the following areas to hatch:

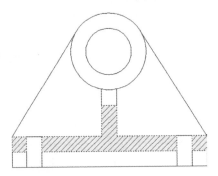

9. Click the Close Hatch Creation button at the right.
10. Start the Hatch command again, and notice that all the options chosen with the previous hatch are still valid. Change the angle to 90°, then press [Enter] to end the command.
11. You will have the following result:

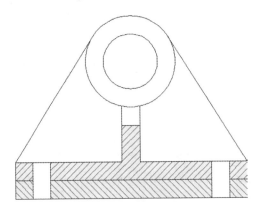

12. Save and close the file.

6.8 SPECIFYING HATCH ORIGIN

- If you want to hatch an area and start the pattern from a certain point and not use the default settings, then you need to manually set the Hatch Origin. By default, AutoCAD uses 0,0 as the starting point for any hatch, which means you will never know for sure if your hatch will be displayed correctly or not. In order to control the Hatch Origin, make sure you are still in the **Hatch Creation** context tab and locate the **Origin** panel. You will see the following:

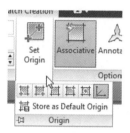

- The first button is **Set Origin**, which will ask you to:

```
Specify origin point:
```

- Specify the desired point. Or you can use the other pre-defined points (lower- left corner, upper-left corner, etc.). You also have the option to save the point you chose this time, for future use, instead of using 0,0.
- See the following example:

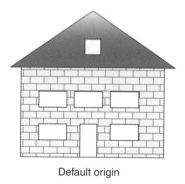

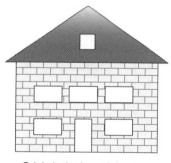

Default origin   Origin is the lower left corner

6.9 CONTROLLING HATCH OPTIONS

- These options control the outcome of the hatching process. Using these options you will be able to hatch an open area (Gap Tolerance), create separate hatches, and more:

6.9.1 How to Use Associative Hatching

- In AutoCAD, hatching is associative, which means the hatch understands the boundary that it fills! When this boundary changes, the hatch will respond appropriately.
- See the illustration below:

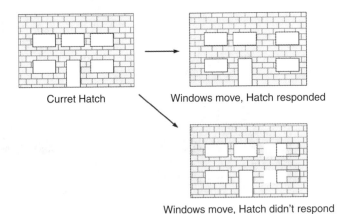

Curret Hatch   Windows move, Hatch responded

Windows move, Hatch didn't respond

6.9.2 How to Make Your Hatch Annotative

- Annotative is an advanced feature related to printing that will be discussed later in this book.

6.9.3 Using Match Properties to Create Identical Hatches

- This option will create an identical hatch from an existing hatch (it will reside in the same layer, and it will have the same angle, scale, transparency, etc.).
- This option has two buttons associated with it: Use current origin or Use source hatch origin. Both options are self-descriptive:

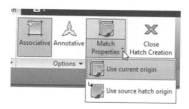

- Select the **Match Properties** button, and you will see the following prompt:

```
Select hatch object:
```

- Select the hatch object you desire to mimic, and you will see the following prompt:

```
Pick internal point or [Select objects/seTtings]:
```

- Click inside the desired area. Keep selecting the area, and when done press [Enter] to end the command.

6.9.4 Hatching an Open Area

- By default, AutoCAD will hatch only closed areas. But you can ask AutoCAD to hatch an area with an opening. To tell AutoCAD to allow hatching an open area, simply set the **Gap Tolerance,** which will be considered the maximum allowable opening. Any area with an opening bigger than this value will not be hatched:

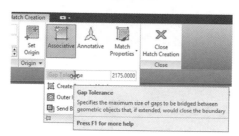

- When you hatch this area, the preview will not be displayed, and you will need to click inside the opened area. You will see the following warning message:

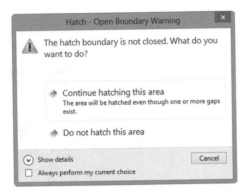

- This message is self-explanatory.

6.9.5 Creating Separate Hatches in the Same Command

- If you are using the same command and you hatched several separate areas, they will be considered a single hatch (single object).
- You can override this default setting by telling AutoCAD that you want separate hatches for separate areas. Simply click this button on:

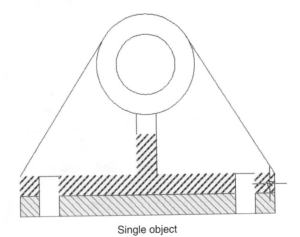

Single object

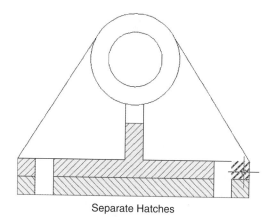

Separate Hatches

6.9.6 Island Detection

- When you are hatching an area containing several areas (islands) inside it, these inside areas may contain more islands. We want to know how AutoCAD will treat these islands.
- There are four different choices:

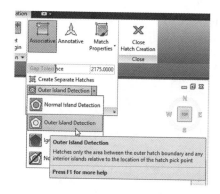

- **Normal Island Detection:** AutoCAD will hatch the first area (the outer one), then leave the second one, hatching the third, and so on.
- **Outer Island Detection:** AutoCAD will hatch the outer area only.
- **Ignore Island Detection:** AutoCAD will ignore all of the inner islands and hatch the outer area, as if there were no areas inside.
- **No Islands Detection:** This option will turn off the Island Detection feature, which will produce the same result as the Ignore Island Detection option.

6.9.7 Set Hatch Draw Order

- You can set the draw order of hatches, relative to the other objects, just like any other object in AutoCAD. You have five choices to choose from:

- These are:
  - Do Not Assign (use the default)
  - Send to Back
  - Bring to Front
  - Send Behind Boundary
  - Bring in Front of Boundary
- See the following example:

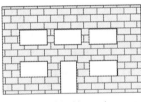
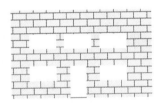

Send behind boundary   Bring in front of boundary

PRACTICE 6-4

Hatch Origin and Options

1. Start AutoCAD 2014.
2. Open **Practice 6-4.dwg**.

3. Start the Hatch command.

4. Using the Options panel, click the Match Properties button, select the hatch of the Toilet, and apply it to the Kitchen; press [Enter] to end the command.

5. Start the Hatch command again, set the background color = 40, and scale = 4, Transparency = 0, and hover over Study room. We need to change the Origin point to the lower-left corner of the room. Click inside Study and end the command.

6. Zoom to the lower-right corner of Living Room, and you will see the area is open. Start the Hatch command again and set the Gap Tolerance = 0.3 (the opening in this drawing is 0.2, so 0.3 is enough), then set to create separate hatches, and click inside Living Room and you will see a warning message. Select the Continue hatching this area option, then click inside Sitting Room and end the command.

7. Thaw layer A-Doors, notice the two doors of the Living Room and Study; they are not shown properly. To solve this problem, select the hatch, right-click, select the Draw Order option, and select the Send to Back option.

8. Start the Hatch command for the fourth time. For Hatch Type, select Solid, and hatch the outside walls.

9. Zoom to any window of the Kitchen, and move it a short distance. What happened to the hatch? _____ Why? _____

10. Save and close the file.

6.10 HATCH BOUNDARIES

- If we were discussing older versions of AutoCAD, this panel (the Boundary panel) would be the first panel to discuss, but because of the instantaneous display of the hatch once you are inside the hatch area, this panel is not as important.

- Depending on what you are doing, creating a new hatch or editing an existing one, some of the buttons will be turned off and some of them will be on:

- The **Pick Points** button is always on, which allows you to choose the areas to hatch, while **Select** and **Remove** allows you to add/remove more objects to be included in the hatch boundaries.
- If you are editing a hatch (by clicking it), the **Recreate** button will be on. This button will help you recreate the boundary if (for any reason) the boundary was deleted. Simply click the hatch without its boundary, then click the **Recreate** button, and you will see the following prompts:

```
Enter type of boundary object [Region/Polyline] <Polyline>:
Reassociate hatch with new boundary? [Yes/No] <N>:
```

- The first prompt asks you to select the desired boundary type, then asks you to re-associate the boundary with the hatch.
- If you select any hatch, AutoCAD will allow you to highlight (**Display**) **Boundary Objects**, so you can edit the boundary. See the following illustration:

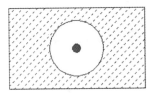   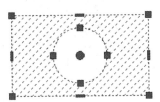

Display boundary objects = Off   Display boundary objects = On

- When you create a hatch, AutoCAD normally creates a polyline (or region) that fits the boundary exactly. Once the command ends, Auto-CAD will delete it. Using the **Retain Boundary** pop-up list you can ask AutoCAD to keep it as a polyline or as a region, or not to keep it at all:

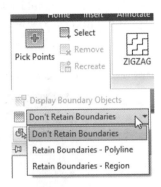

- In order for the Hatch command to work, it needs to analyze all the objects in the current viewport (in Model space this means the area you are seeing right now), which may take a long time (depending on the number of objects), but you can ask AutoCAD to analyze only the relative objects rather than all objects. Locate the **Select New Boundary Set** button:

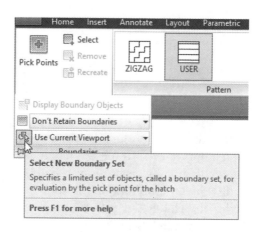

- You will see the following prompt:

```
Select objects:
```

- Select the desired objects, then press [Enter], and this time the pop-up list will read **Use Boundary Set**, instead of the default prompt:

6.11 EDITING HATCHES

- Editing hatches in AutoCAD has never been easier. There are two ways to edit a hatch: by clicking (single-click) or by using Properties (double-click).
- If you single-click a hatch, three things will take place:
  - You will see a grip (small dot) at the center of the area.
  - The context **Hatch Editor** tab will appear, which includes the same panels the Hatch Creation tab does.
  - The Quick Properties palette will appear.

- If you move your mouse to the grip (without clicking), you will see the following menu:

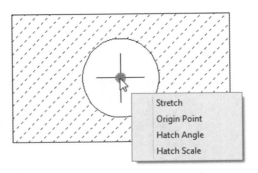

- You can do any or all of the following:
  - Stretch in order to stretch the hatch; however, it is preferable to display boundary objects as discussed above.
  - Modify the Origin Point on the spot.
  - Modify the Hatch Angle on the spot.
  - Modify the Hatch Scale on the spot.
- The context tab Hatch Editor also allows you to make all the necessary modifications because it contains the same panels as the Hatch Creation tab.
- The **Quick Properties** palette will appear as well, and you can make any edits needed:

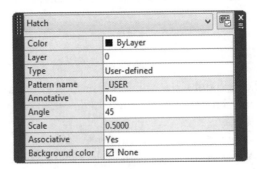

- On the other hand, double-clicking the desired hatch, or selecting the hatch, right-clicking, and selecting the **Properties** option will show the **Properties** palette:

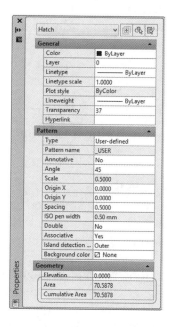

- The data available for editing is everything related to the hatch properties, options, and boundaries. The Properties palette also provides a single piece of information that other methods don't: the Area of the hatch. If you select a single hatch (whether a single area or multiple areas hatched as a single area) the **Area** field and the **Cumulative Area** will be the same. But if you select multiple areas created using different commands, only the **Cumulative Area** will be filled and the **Area** field showing **Varies** will have a value.

- If you select the hatch and right-click, you will see some Hatch-related editing commands:

- These commands include:
  - **Hatch Edit**, which will show the old Hatch dialog box.
  - **Set Origin**, which will help you set a new origin.
  - **Set Boundary**, which will help you set a new boundary set
  - **Generate Boundary**, which will regenerate a new boundary for this hatch.

PRACTICE 6-5

Hatch Boundary and Hatch Editing

1. Start AutoCAD 2014.
2. Open **Practice 6-5.dwg**.
3. Select the existing hatch at the top left of the drawing.
4. From the Boundaries panel, click Recreate, then select **Polyline**, and Yes. You can see that AutoCAD recreated a new boundary for this orphan hatch.
5. Using the existing hatch at the middle left, select the hatch so you can see the special grip, and then change the angle to 90°.
6. Using the existing hatch at the bottom left, select the hatch, so you can see the special grip, and then change the origin point to the lower-left corner and the scale to = 0.1.
7. Using the existing hatch at the upper right, select it and right-click, then select the Properties option; change the scale to 0.5 and the background color to Magenta. What is the total area of the hatch? _____
8. Start the Hatch command, select pattern name = steel, and set the scale = 2.0. Now go to the Options panel, and change Island detection to Normal. Try Hatching the shape at the middle right. Change the Island detection to Outer, and try it, then Ignore and try it. Finally, get it back to Normal and finish the command.
9. Save and close the file.

NOTES:

CHAPTER REVIEW

1. The command to convert a block to a file is:
   a. Makeblock
   b. Createblock
   c. Wblock
   d. None of the above
2. There are _____ hatch types in AutoCAD.
3. Using Tool Palettes you can customize all tools by changing their properties.
   a. True
   b. False
4. Hatch grips are like regular grips.
   a. True
   b. False
5. If both Design Center and Tool Palettes are open, you can drag any block from Design Center to a tool Palette.
   a. True
   b. False

CHAPTER REVIEW ANSWERS

1. c
3. a
5. a

7 WRITING TEXT

In This Chapter

◇ How to insert single line and multiline text
◇ How to edit text

7.1 INSERTING SINGLE LINE TEXT

- This command allows you to create single lines of text, and each line will be independent of the other lines. To issue this command, go to the **Annotate** tab, locate the **Text** panel, and then click the **Single Line** button:

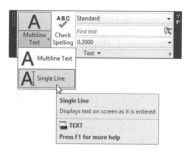

- You will see the following prompts:

```
Current text style: "Standard" Text height: 0.2000
Annotative: No Justify: Left
Specify start point of text or [Justify/Style]:
Specify rotation angle of text <0>:
```

- The first prompt gives you some information about the current settings, the current style (in our example Standard), the current height (in our example = 0.20), whether this text will be Annotative or not, and the justification of the text.
- At the first prompt you can change the current Justification (discussed in the section on Multiline Text) and Style settings by typing **J** or **S**, or you can specify the start point of the baseline of the text, then specify the rotation angle (default value is 0 (zero)). Once you press [Enter] you will see a blinking cursor ready for you to type any text you want. To finish any line, press [Enter], and to end the command press [Enter] twice.
- Another way of setting the current text style is to go to the **Annotate** tab, then locate the **Text** panel, and you will you see at the top-right the current text style, which you can change:

PRACTICE 7-1

Creating a Text Style and Inserting Single Line Text

1. Start AutoCAD 2014.
2. Open **Practice 7-1.dwg**.
3. Make the Room Names text style current.
4. Make layer Text current.
5. Enter the room names as shown below.
6. Make the Standard text style current (with this text style the height = 0, so you should set it every time you want to use this style).
7. Make layer Centerlines current.
8. Zoom to the upper-left centerline and notice that the letter A is missing.
9. Start Single Line text, right-click, and select the Justify option. From the list choose MC (Middle Center), select the center of the circle as the Start point, set the Height to 0.25 and the Rotation angle = 0, then type A, and press [Enter] twice.

10. You should have the following:

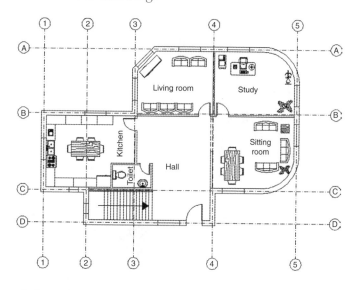

11. Save and close the file.

7.2 INSERTING MULTILINE TEXT

- This command allows you to type text in an environment similar to Microsoft Word or other word processor. To issue this command, go to the **Annotate** tab, locate the **Text** panel, and then click the **Multiline Text** button:

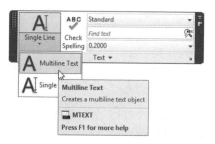

- You will see something like the following in the Command window:

```
Current text style: "TNR_05" Text height: 0.500
Annotative No
Specify first corner:
Specify opposite corner or [Height/Justify/Line
spacing/Rotation/Style/Width]:
```

- The first prompt gives you some information about the current settings, including the current style (in our example TNR_05) and the current height (in our example = 0.50) and whether this text will be Annotative or not. You will need to specify an area to write in, so the cursor will change to:

- At the Command window, you will see the following prompt:

```
Specify first corner:
```

- Specify the first corner, and you will see something like the following:

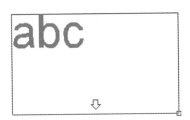

- At the Command window, you will see the following prompt:

```
Specify opposite corner or [Height/Justify/Line spacing/
Rotation/Style/Width/ Columns]:
```

- Specify the other corner, or select one of the options, and the Text Editor with a ruler will appear:

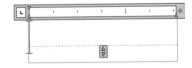

- A context tab called **Text Editor** will appear, which looks like the following:

- These panels allow you to do many things; they are discussed as follows.

7.2.1 Style Panel

- This is the Style panel:

- Use the Style panel to select the text style you want to use and set the height (this value will overwrite the text style height, so be careful).

7.2.2 Formatting Panel

- This is the Formatting panel:

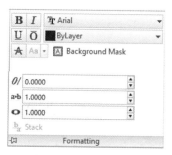

- Use the **Formatting** panel to do all or any of the following:
  - Make the selected text **Bold**
  - Make the selected text **Italic**
  - Make the selected text **Underlined**
  - Make the selected text **Overlined**
  - Make the selected text **Strikethrough**
  - Change the **Font** of the selected text
  - Change the **Color** of the selected text

- Change capital letters to lowercase letters and vice-versa
- Change the **Background Mask** for the selected text, and you will see the following dialog box:

- Change the **Oblique Angle** of the selected text
- Change the **Tracking** (to increase or decrease the spaces between letters; if the value is greater than 1 there will be more spaces between letters and vice-versa)
- Change the **Width Factor**

7.2.3 Paragraph Panel

- This is the Paragraph panel:

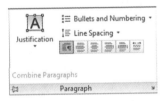

- Use the Paragraph panel to do all or any of the following:
  - Change the **Justification** of the text relative to the text area selected; choose one of the following options:

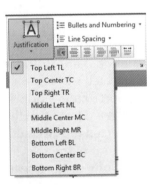

- The following is an illustration of the nine points available relative to the text area:

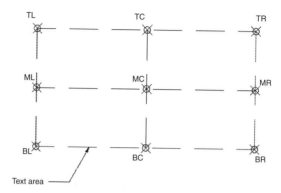

- Change the text selected to use **Bullets and Numbering**; there are three choices (Numbered, Lettered, Bulleted):

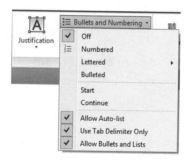

- Change the **Line Spacing** of the paragraph:

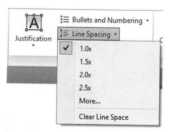

- Change the horizontal justification of the text by using one of the six buttons shown below:

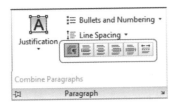

7.2.4 Insert Panel

- This is the Insert panel:

- In the Insert panel, you can do all or any of the following to the selected text:
  - Convert text to two **Columns** or more. If you click the **Columns** button, the following menu will appear:

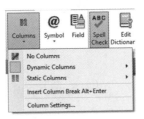

  - Use **Dynamic Columns** to select whether you want AutoCAD to specify the height (Auto height) or if you want to set the height (Manual height):

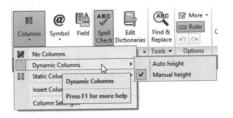

- Select the **Static Columns** option to specify the number of columns:

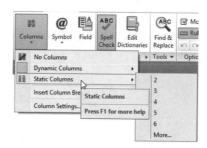

- Select the **Insert Column Break Alt+Enter** option to insert a column break at a certain line, which means the rest of the column will go to the next column.
- Select the **Column Settings** option to show the **Column Settings** dialog box, which allows you to adjust all of the above settings as well as set the **Column** width and **Gutter** distance:

- Here is an example:

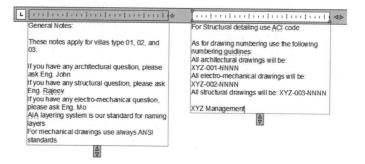

- Select **Symbol** to add scientific characters to your text. You will see the 20 available symbols:

7.2.5 Spell Check Panel

- This is the Spell Check Panel:

- As long as you are in the Text Editor, you can leave the **Spell Check** button on to catch any misspelled words. You will see a dotted red line underneath the misspelled words, as shown in the following example:

- To get suggestions for the correct word, move to the word, right-click, and you will see something like the following:

- Select the **Edit Dictionaries** button and you can select another dictionary other than the default one.

7.2.6 Tools Panel

- This is the Tools panel:

- In the **Tools** panel, you can do all or any of the following:
  - Click the **Find & Replace** button to search for a word and then replace it with another word. You will see the following dialog box:

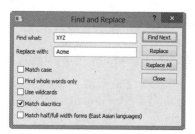

  - Use the **Import Text** button to import text from a text file.
  - Use the **AutoCAPS** button to type in the Text Editor using capital.

7.2.7 Options Panel

- This is the Options panel:

- Using the Options panel you can do all or any of the following:
  - Change the **Character Set**, change the **Editor Settings**, or learn more about multiline text through **Help**:

  - Show the ruler (by default it is displayed) or hide it
  - Undo and redo text actions

7.2.8 Close Panel

- The following shows the Close panel, which contains a single button to close the Text Editor and all its panels:

7.2.9 While You Are in the Text Editor

- While you are in the Text Editor, you can do the following things:
  - Using the ruler you can set the First Line indent and the Paragraph indent:

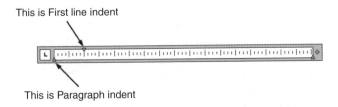

- If you right-click you will see the following menu, which includes all the functions discussed above:

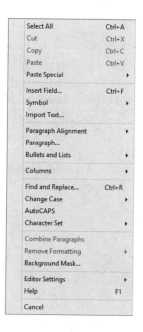

- You can change the width and height of the area using the following controls:

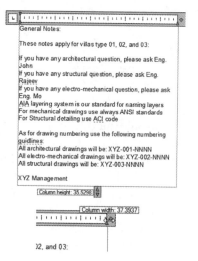

PRACTICE 7-2

Multiline Text

1. Start AutoCAD 2014.
2. Open **Practice 7-2.dwg**.
3. You are now in the layout called "Cover."
4. Make sure that the current layer is Text.
5. Make the text style Title is the current text style.
6. Start Multiline Text, and specify the two corners of the rectangle as your text area.
7. Before you type anything, change the Justification to MC.
8. Type the following as shown:

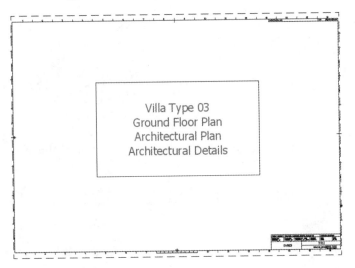

9. Erase the rectangle.
10. Go to the layout called ISO A1 – Overall.
11. Make the Notes text style the current text style.
12. Using the rectangle at the right, specify the two corners of your text area.
13. Using the Tools panel, select the Import Text command, and select the *Notes.text* file.
14. Select General Notes and make it Bold, Underlined, and size = 5.0.
15. Stretch the width a little bit so 03 will be on the same line.

16. Select the text from the first line, "If you have ...." to "ACI code," then select Bullets and Numbering and select Numbered.
17. Select the text starting from, "As for ...." until the last "NNNN," and make it Bulleted and Line spacing = 1.5 x.
18. Select the last line "XYZ Management" and make it Centered and Italic.
19. Close the Text Editor.
20. Erase the rectangle.
21. Go to ISO A1 - Architectural Details.
22. Using the rectangle, start Multiline Text, select two opposite corners, and import the same file, *Notes.txt*.
23. Do the same thing you did above and make the text Numbered and Bulleted.
24. Start the Column command and set two static columns; then using the Column Settings dialog box, set Column Width = 125 and Gutter = 22.5.
25. Using the two arrows at the lower-left corner, make sure that all six points are in the first column.
26. Erase the rectangle.
27. This what you should have:

General Notes:

These notes apply for villas type 01, 02, and 03:

1. If you have any architectural question, please ask Eng. John
2. If you have any structural question, please ask Eng. Rajeev
3. If you have any electro-mechanical question, please ask Eng. Mo
4. AIA layering system is our standard for naming layers
5. For mechanical drawings use always ANSI standards
6. For Structural detailing use ACI code

- As for drawing numbering use the following numbering guidlines:
- All architectural drawings will be: XYZ-001-NNNN
- All electro-mechanical drawings will be: XYZ-002-NNNN
- All structural drawings will be: XYZ-003-NNNN

XYZ Management

28. Save and close the file.

7.3 TEXT EDITING

- There are several ways to edit text including double-clicking and using Quick Properties, Properties, and Grips.

7.3.1 Double-Clicking Text

- To edit the text, single line or multiline, simply double-click it. If it is single line text, you will see the text selected and be able to add to it or modify it. If it is multiline text, you will see the Text Editor again and the Text Edit context tab will appear at the top where you can make all your changes.

7.3.2 Quick Properties and Properties

- To show the Quick Properties for either single line text or multiline text, simply click the desired text. You will see something like the following (Text in the following illustration is single line text and MText is multiline text):

- You can change the following settings: Layer, Contents, Style, Annotative (Yes, or No), Justification, Height, and Rotation angle.
- If you select single line text or multiline text and right-click, then select the Properties option, you will see something like the following (again, Text is single line text and MText is multiline text):

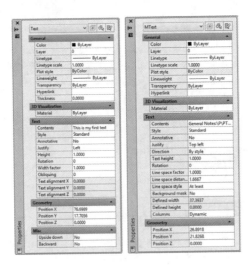

- As you can see, you can change anything related to the selected text.

7.3.3 Editing Using Grips

- You will see a grip at the start point of the baseline and another one at the Justification point (you may see only one point if both points coincide) if you click single line text:

▪This is my first text

- You will see a single grip appear at the Justification point selected when the multiline text is selected and then you will see two triangles: one at the lower part of the text and one at the right side. The lower part triangle allows you to cut your multiline text into columns; simply stretch it up and the text will be cut into two columns. The triangle on the right allows you to increase/decrease the horizontal distance of the text, hence increasing/decreasing the height of the text:

▪These notes apply for villas type 01, 02, and 03: ▶

▼

- If you click any multi-column text, you will see the same thing for the first column at the left, except the arrow at the bottom will be pointing downward instead of upward. For the other columns you will see an arrow at the right side. For the last column on the right, you will see an extra arrow at the right, which allows you to increase/decrease the width and height of the whole text in one step. See the following illustration:

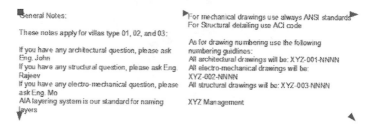

7.4 SPELL CHECK AND FIND AND REPLACE

- While you are in the Text Editor, you can use Spell Check and Find and Replace, but what if you aren't in the Text Editor? You can still use these features. AutoCAD can spell check the whole drawing and not just the current text; AutoCAD can also spell check the current space and/ or layout (layouts will be discussed in Chapter 9) or any selected text. To issue the Check Spelling command, go to the **Annotate** tab, locate the **Text** panel, and click the **Check Spelling** button:

- You will see the following dialog box:

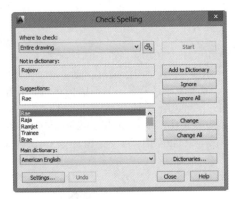

- If AutoCAD finds any misspelled word, it will give you suggestions to choose from or you can simply ignore the misspelled word altogether. AutoCAD can also find any word or part of a word and replace it in the entire drawing file.

- To issue the **Find and Replace** command, go to the **Annotate** tab, locate the **Text** panel, and then type the word you want to replace in the *Find text* field, as shown below, then click the small button at the right:

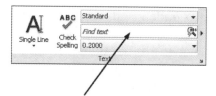

Type here the text that you
want to find and replace

- You will see the following dialog box:

- Under **Replace with**, type the new word(s) you want, and then select one of the following choices: **Find**, **Replace**, and **Replace All**. When done, click **Done**.

PRACTICE 7-3

 Editing Text

1. Start AutoCAD 2014.
2. Open **Practice 7-3.dwg**.
3. You are now in the Cover layout. Click the text, using the arrow at the bottom, and drag it upward to cut the text into two columns, each holding two lines.

4. Use the arrow at the right of the second column and drag it until the two columns touch each other. Press [Esc] to end the editing process.
5. Go to layout ISO A1 – Overall.
6. Zoom to the text at the right, select it, right-click and choose the Properties option, and change the following:
   a. Change the style from Notes to Standard
   b. Change Justify to Middle Center
   c. Change Text Height to 4.5
7. Using the arrow at the right, stretch the text to the right by 10 units (if OSNAP is annoying you turn it off).
8. Press [Esc] to end the editing process.
9. Go to the Annotate tab, locate the Text panel, and in the Find and Replace field, type XYZ, and then click the small button at the right. When the dialog box comes up, in the Replace with field, type ACME, and then click Replace All.
10. Go to layout ISO A1 Architectural Details, double-click the multi-column text, and convert it to single column text.
11. Press [Esc] to end the editing process.
12. Go to Model space.
13. Click the word "Hall," and the Quick Properties will appear. Set the Justification to Middle Center. Using the Move command try to put this word in the middle center of the space.
14. Save and close the file.

NOTES:

CHAPTER REVIEW

1. There are two types of text in AutoCAD, Single Line and Multiline.
   a. True
   b. False
2. Which one of the following is *not* a panel in the Text Editor context tab?
   a. Insert
   b. Options
   c. Text
   d. Paragraph
3. When you start the Single Line text command, AutoCAD will tell you the current text style.
   a. True
   b. False
4. While you are in the Multiline text command, select the _____ _____ option to make the rest of the column go to the next column.
5. In the Multiline text command, you can set the background for a text.
   a. True
   b. False
6. Degrees, Centerline, and Not Equal are _____ you can insert in multiline text command.

CHAPTER REVIEW ANSWERS

1. a
3. a
5. a

Chapter **8** DIMENSIONS

In This Chapter

◇ Dimensioning, and dimension types
◇ How to insert different types of dimensions
◇ How to edit dimension blocks

8.1 WHAT IS DIMENSIONING?

■ Dimensioning in AutoCAD is just like using text and tables. You should create your dimension style first, and then use it to insert dimensions. Dimension styles control the overall outcome of the dimension block generated by the different types of dimension commands.

■ To insert a dimension, depending on the type of the dimension, you should specify points, or select objects, and then a dimension block will be added to the drawing. For example, in order to add a linear dimension you will select two points representing the distance to be measured, and a third point will be the location of the dimension block. See the illustration below:

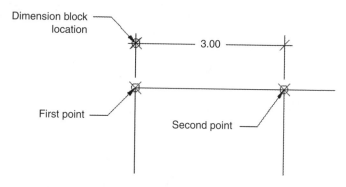

- The generated block consists of three things:
    - Dimension line
    - Extension lines
    - Dimension text
- See the following illustration:

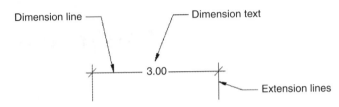

8.2 DIMENSION TYPES

- The following are the dimensions types in AutoCAD:

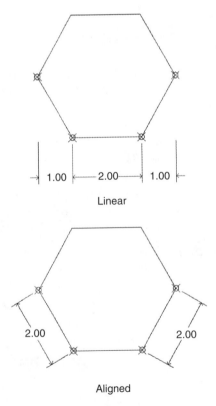

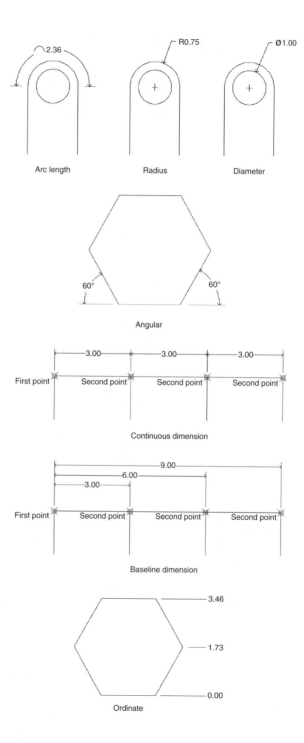

Arc length   Radius   Diameter

Angular

Continuous dimension

Baseline dimension

Ordinate

8.3 HOW TO INSERT A LINEAR DIMENSION

- ■ This command allows you to create a horizontal or vertical dimension. To start the Linear command, go to the **Annotate** tab, locate the **Dimensions** panel, and then select the **Linear** button:

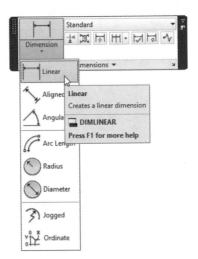

- ■ You will see the following prompts:

```
Specify first extension line origin or <select object>:
Specify second extension line origin:
Specify dimension line location or
[Mtext/Text/Angle/Horizontal/Vertical/Rotated]:
```

- ■ Specify the first point and second point of the dimension distance to be measured and then specify the location of the dimension block by specifying the location of the dimension line. The following is the result:

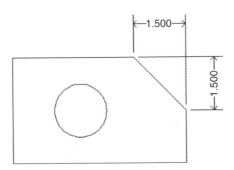

- The prompts also contain other options:
  - Mtext
  - Text
  - Angle
  - Horizontal
  - Vertical
  - Rotated
- **Mtext** allows you to edit the measured distance in MTEXT mode, while **Text** allows you to edit the measured distance in DTEXT (Single line) mode. **Angle** is used to change the angle of the text (the default is 0 (zero)). **Horizontal** and **Vertical** will force the dimension to be either horizontal or vertical (the default allows you to specify either using your mouse). Finally, **Rotated** is used to create a dimension line parallel to another angle given by you.

8.4 HOW TO INSERT AN ALIGNED DIMENSION

- This command allows you to create a dimension parallel to the two points specified. To start this command, go to the **Annotate** tab, locate the **Dimensions** panel, and then select the **Aligned** button:

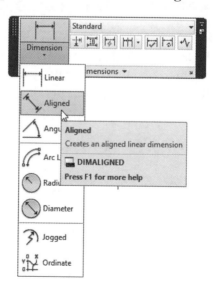

- The following prompts will appear:

```
Specify first extension line origin or <select object>:
Specify second extension line origin:
Specify dimension line location or
[Mtext/Text/Angle]:
```

- Specify the first point and second point of the dimension distance to be measured and then specify the location of the dimension block by specifying the location of the dimension line. See the following illustration:

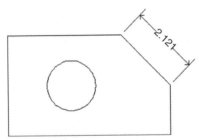

- The rest of the options are identical to the **Linear** command prompts.

8.5 HOW TO INSERT AN ANGULAR DIMENSION

- This command allows you to insert an angular dimension between two lines, the included angle of a circular arc, the two points and center of a circle, or three points. To start this command, go to the **Annotate** tab, locate the **Dimensions** panel, and then select the **Angular** button:

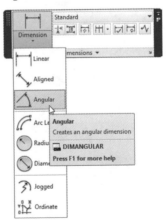

- AutoCAD may use one of the following methods based on the selected objects:
  - If you select a circular arc, it will measure the included angle.
  - If you select a circle, your selecting points will be the first point, the center of the circle will be the second point, and then you will select third point.
  - If you select a line, it will ask you to select a second line.
  - If you select a point, it will be considered the center point, and it will ask you to specify two more points.
- See the following illustration:

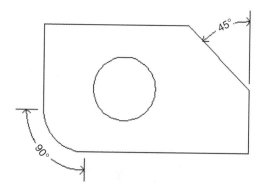

- Based on the above discussion, when you start the command you will see the following prompts (in the following example we selected an arc):

```
Select arc, circle, line, or <specify vertex>:
Specify dimension arc line location or [Mtext/Text/Angle]:
```

PRACTICE 8-1

Inserting Linear, Aligned, and Angular Dimensions

1. Start AutoCAD 2014.
2. Open **Practice 8-1.dwg**.
3. Make layer Dimension the current layer.

4. Insert the dimensions as shown below:

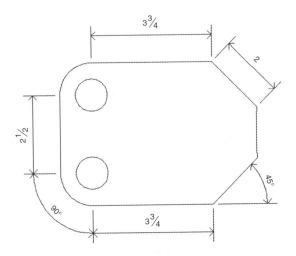

5. Save and close the file.

8.6 HOW TO INSERT AN ARC LENGTH DIMENSION

- This command allows you to create a dimension measuring the length of an arc. To start this command, go to the **Annotate** tab, locate the **Dimensions** panel, and then select the **Arc Length** button:

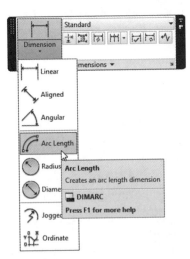

- You will see the following prompts:

```
Select arc or polyline arc segment:
Specify arc length dimension location, or [Mtext/Text/
Angle/Partial/Leader]:
```

- Select the desired arc and then locate the dimension block, either inside or outside the arc. You will get something like the following:

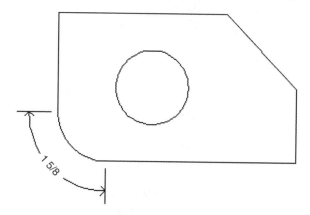

- The options, Mtext, Text, and Angle, were already discussed previously. The **Partial** option means you want to insert an arc length dimension on part of the arc. You select the arc, then select two internal points on the arc, and you will get something like the following:

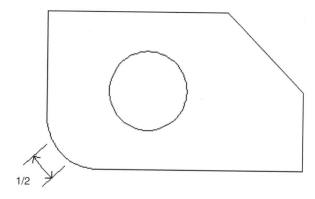

8.7 HOW TO INSERT A RADIUS DIMENSION

- This command allows you to insert a Radius dimension on an arc or circle. To start this command, go to the **Annotate** tab, locate the **Dimensions** panel, and then select the **Radius** button:

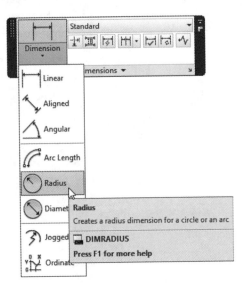

- You will see the following prompts:

```
Select arc or circle:
Specify dimension line location or [Mtext/Text/ Angle]:
```

- Select the desired arc or circle and then locate the dimension block. You will get something like the following:

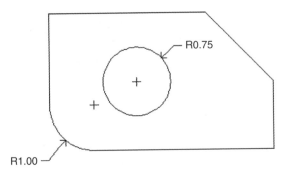

8.8 HOW TO INSERT A DIAMETER DIMENSION

- This command allows you to insert a diameter dimension on an arc or circle. To start this command, go to the **Annotate** tab, locate the **Dimensions** panel, and then select the **Diameter** button:

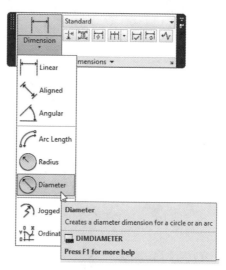

- You will see the following prompts:

```
Select arc or circle:
Specify dimension line location or [Mtext/Text/ Angle]:
```

- Select the desired arc or circle and then locate the dimension block. You will get something like the following:

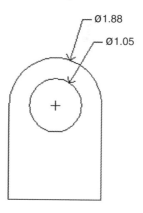

PRACTICE 8-2

Inserting Arc Length, Radius, and Diameter Dimensions

1. Start AutoCAD 2014.
2. Open **Practice 8-2.dwg**.
3. Make layer Dimension the current layer.
4. Insert the dimensions as shown below:

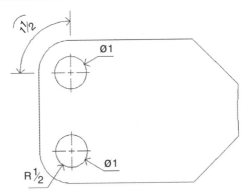

5. Save and close the file.

8.9 HOW TO INSERT A JOGGED DIMENSION

- This command allows you to insert a jogged arc dimension for a big arc, simulating a new center point. To start this command, go to the **Annotate** tab, locate the **Dimensions** panel, and select the **Jogged** button:

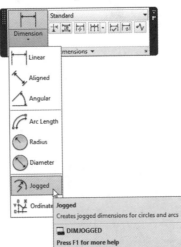

- The following prompts will appear:

```
Select arc or circle:
Specify center location override:
Dimension text = 1.5
Specify dimension line location or [Mtext/Text/ Angle]:
Specify jog location:
```

- As a first step, select an arc or circle, specify the point that will be the new center point (AutoCAD calls it location override), locate the dimension line, and finally, specify the jog's location. You will get something like the following:

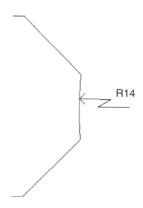

8.10 HOW TO INSERT AN ORDINATE DIMENSION

- This command allows you to insert dimensions relative to a datum, either in X or in Y. See the following illustration:

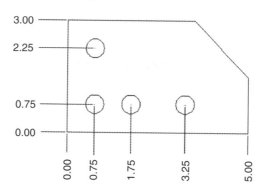

NOTE ➤ ■ Use the UCS command and the Origin option to relocate the origin to one side of the shape, so the values in both X, Y will be correct. If you leave the origin as the current UCS origin, the values inserted may be incorrect.

■ To start this command, go to the **Annotate** tab, locate the **Dimensions** panel, and then select the **Ordinate** button:

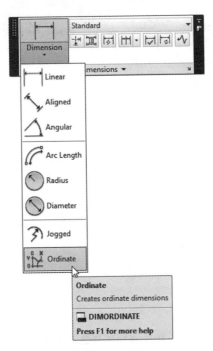

■ You will see the following prompts:

```
Specify feature location:
Specify leader endpoint or [Xdatum/Ydatum/Mtext/Text/
Angle]:
```

■ First, select the desired point. By default, AutoCAD will give you the freedom to go in the direction of X or in Y. If you want to force the mouse to measure points related to the X axis, then select the **Xdatum** option; the same applies for the Y axis. We have already discussed the rest of the options.

PRACTICE 8-3

Inserting Dimensions

1. Start AutoCAD 2014.
2. Open **Practice 8-3.dwg**.
3. Make layer Dimension current.
4. Insert the dimensions as shown below:

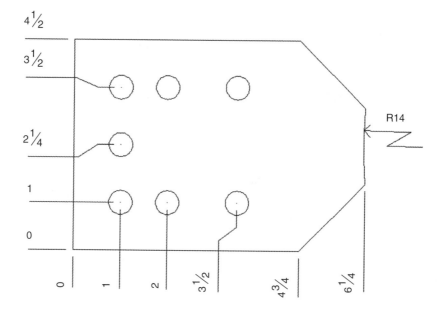

5. Save and close the file.

8.11 INSERTING A SERIES OF DIMENSIONS USING THE CONTINUE COMMAND

- AutoCAD allows you to input series of dimensions using the Continue command. The Continue command will follow the last dimension command by asking you to input the second point, assuming that the last point of the last dimension will be considered the first point. To start this command, go to the **Annotate** tab, locate the **Dimensions** panel, and then select the **Continue** button:

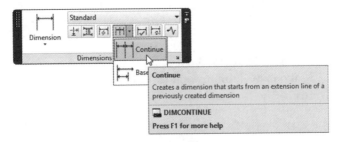

- There are two scenarios when using the Continue command:
  - If there wasn't a dimension command issued in the current AutoCAD session, AutoCAD will ask you to select an existing dimension (linear, ordinate, or angular). You will see the following prompt:

  ```
  Select continued dimension:
  ```

  - If there was a dimension command issued in the current AutoCAD session, AutoCAD will ask you to continue this command by asking you to specify the second point. You can also select an existing dimension block to continue it, or you can undo the last continue command. You will see the following prompt:

  ```
  Specify a second extension line origin or [Undo/
  Select] <Select>:
  ```

  - You will get something like the following:

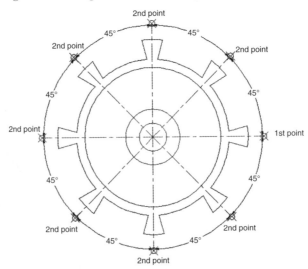

8.12 INSERTING A SERIES OF DIMENSIONS USING THE BASELINE COMMAND

- This command is identical to the Continue command, except all dimensions will be measured based on the first point specified by you as the baseline. To start this command, go to the **Annotate** tab, locate the **Dimensions** panel, and then select the **Baseline** button:

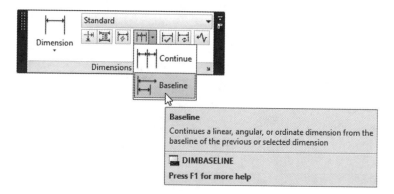

- We don't need to discuss the prompts of this command because they resemble the Continue command prompts. You will get something like the following:

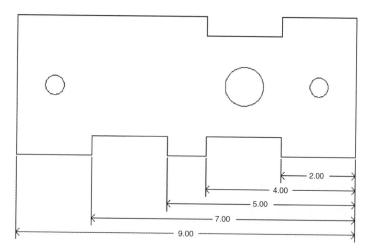

PRACTICE 8-4

Continue Command

1. Start AutoCAD 2014.
2. Open **Practice 8-4.dwg**.
3. Create an angular dimension, and then using the Continue command, complete the shape as follows:

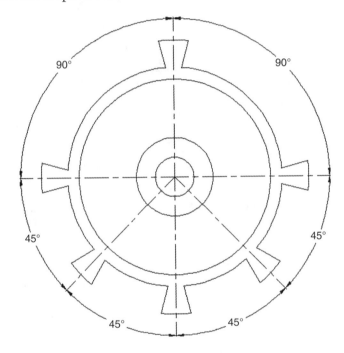

4. Save and close the file.

PRACTICE 8-5

Baseline Command

1. Start AutoCAD 2014.
2. Open **Practice 8-5.dwg**.

3. Create a linear dimension, and then using the Baseline command, complete the shape as follows:

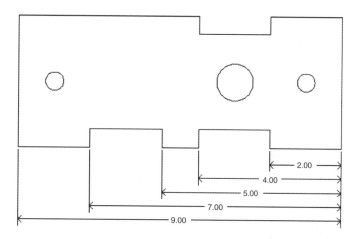

4. Save and close the file.

8.13 USING THE QUICK DIMENSION COMMAND

- This command allows you to insert a group of dimensions in one step. To start this command, go to the **Annotate** tab, locate the **Dimensions** panel, and then select the **Quick Dimension** button:

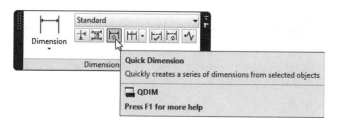

- You will see the following prompt:

```
Associative dimension priority = Endpoint
Select geometry to dimension:
Specify dimension line position, or [Continuous/ Staggered/
Baseline/Ordinate/Radius/ Diameter/ datumPoint/Edit/
seTtings] <Continuous>:
```

- As a first step, select the geometry you want to dimension using clicking, Window mode, Crossing mode, or any other mode you know. If this is the first time, you are using the command in the current AutoCAD session, then AutoCAD will use Continuous as the default option. But if you right-click, you will see the following menu:

- Using this shortcut menu, select the desired dimension type and then specify the dimension line location; as a result, a set of dimensions will be inserted.
- See the following examples:

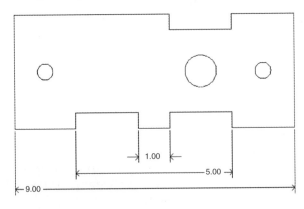

Staggered

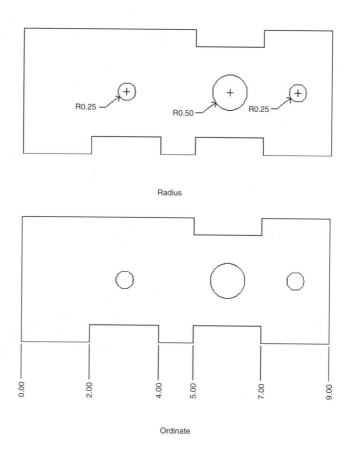

Radius

Ordinate

- If you select the **Settings** option, you will see the following menu:

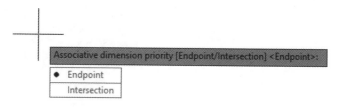

- This option will allow you to set the default OSNAP for specifying extension line origins.

8.14 EDITING DIMENSION BLOCK USING GRIPS

- After inserting a dimension block it is easy to edit it using grips, or by right-clicking. Depending on the type of dimension, if you click any dimension block you will see grips in certain places. The following are some examples:
  - For linear and aligned dimensions, grips will appear in five places: at the two points measured, the two ends of the dimension line, and finally, at the dimension text. If you *hover* over the text grip, you will see the following:

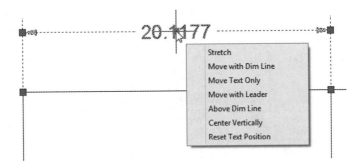

  - AutoCAD allows you to Stretch, Move with Dim Line, Move Text Only, etc., while holding this grip. If you hover over the two grips at the ends, you will see:

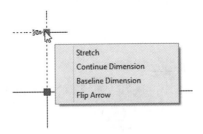

  - Here, you can Stretch and create a continuous or baseline dimension based on the type of dimension selected. You can also flip the arrow nearest to the selected grip.
  - Angular dimension grips will appear at five places: at the end points of the two lines involved, at the dimension line, and finally,

at the dimension text. If you hover over the text grip, you will see the following:

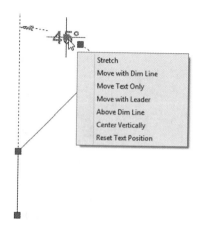

- The commands are identical to linear and aligned dimensions. If you hover over the two end grips, you will see the following:

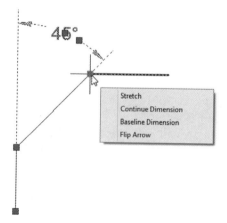

- These are the same commands discussed for linear and aligned dimensions.
- Ordinate dimension grips will appear at four places: the origin point and the measured point, then at the dimension line, and finally, at

the dimension text. If you hover over the text grip, you will see the following:

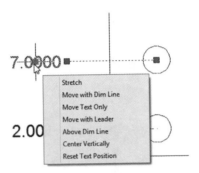

- These are the same commands discussed above. Hovering over the end of the line grip will show the following:

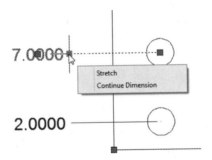

- Ordinate will not work with Baseline, and doesn't have arrows, so these two options are not given for this type of dimension.
- Radius and diameter dimension grips will appear at three places: at the selected point, at the center, and finally, at the dimension text. If you hover over the text grip, you will see the following:

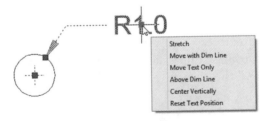

- This the same list of commands. If you hover over the grip at the end of the arrow, you will see the following:

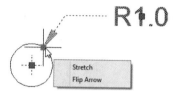

- Since radius and diameter don't have the ability to use Continue or Baselines, these two commands are not mentioned. You can use only Stretch and Flip Arrow.
- Arc length will show four grips: one at the two ends of the arc, one at the text, and finally, one near the text. If you hover over the text grip, you will see the following:

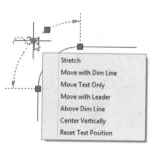

- This is also the same list of commands. If you hover over the end point grip, you will see the following:

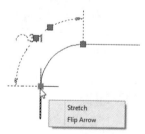

- Since arc length will not work with Continue and Baseline, only Stretch and Flip Arrow will appear.

8.15 EDITING DIMENSION BLOCK USING THE RIGHT-CLICK MENU

- On the other hand, if you select a dimension block and right-click, you will see the following shortcut menu:

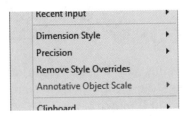

- You can change the dimension style used to insert the selected dimension block. Or better yet, you can save the changes you made as a new dimension style, as shown below:

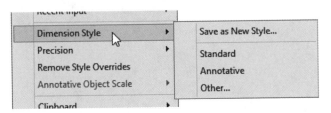

- You can also change the Precision of the dimension text:

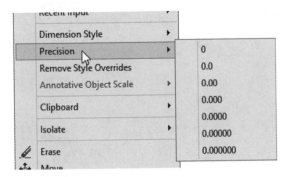

8.16 EDITING A DIMENSION BLOCK USING QUICK PROPERTIES AND PROPERTIES

- If you select a dimension block, Quick Properties will come up automatically. You will see something like the following:

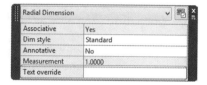

- You have the ability to modify the dimension style and choose the Annotative feature (which will be discussed in the next chapter). You will also find the exact measurement of the selected dimension, and you can change it using the Text override field.
- Properties, on the other hand, allows you to make global changes to the selected dimension blocks. Simply select the dimension block(s), right-click, and then select the **Properties** option (you can double-click as well). You will see something like the following:

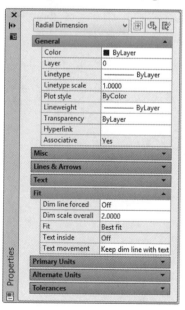

- Using the Properties palette you can change everything related to the dimension block or the dimension style of the block (in the above example we showed the variables you can control under the Fit category).

PRACTICE 8-6

Quick Dimensions and Editing

1. Start AutoCAD 2014.
2. Open **Practice 8-7.dwg**.
3. Input the following dimensions using all the techniques you have learned:

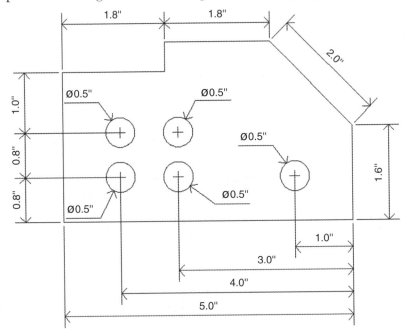

4. Using all the editing techniques you have learned make the following changes:

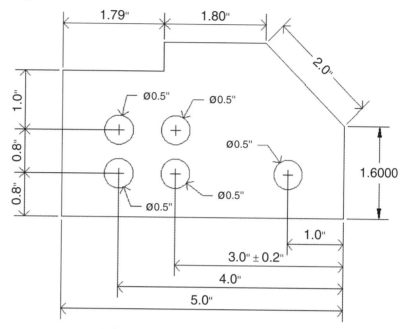

5. In order to set the Tolerance for the 3" dimension use Properties, then Tolerance, and set the Tolerance type to Symmetrical and the value to be 0.2.
6. Save and close the file.

NOTES:

CHAPTER REVIEW

1. The Arc length command should be used only with arcs and polylines, not with circles.
   a. True
   b. False
2. Continue and Baseline can't work with which of the following?
   a. Linear
   b. Radius
   c. Ordinate
   d. Angular
3. _____ allows you to insert dimensions relative to a datum, either in X or in Y.
4. Inputs to use this dimensioning command may be three points, or two lines:
   a. Linear
   b. Radius
   c. Ordinate
   d. Angular
5. Which one of the following is not among the commands of Quick Dimension?
   a. Continuous
   b. Radius
   c. Staggered
   d. Angular

CHAPTER REVIEW ANSWERS

1. a
3. Ordinate
5. d

9 PLOTTING

In This Chapter

◇ The difference between Model Space and Paper Space
◇ How to create a new layout using different methods
◇ How to create and control viewports

9.1 WHAT ARE MODEL SPACE AND PAPER SPACE?

- AutoCAD provides two working spaces; one for creating your drawing, called Model space, and one for plotting your drawing, called Paper space. There is only one Model space in each drawing, but there are an infinite number of Paper spaces per file, and each one is called a layout.
- Each layout is linked to a page setup where you specify everything related to plotting including the plotter, paper size, and paper orientation (Portrait or Landscape).
- After you create the layout and link it to a page setup, you will insert a title block and then add viewports (which represent a portion of the Model space). You will also set up the scale of the viewports.

9.2 INTRODUCTION TO LAYOUTS

- The layout is the place you will plot your drawing from. Each layout will be linked to a **Page Setup**, **Objects**, such as the title block, text, dimensions, and **Viewports**, which will be covered separately in the coming discussion. See the roadmap illustrated below:

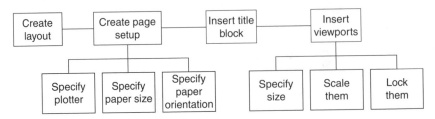

- Each layout should have a name. By default when you create a new drawing using the *acad.dwt* template, two layouts will be included: Layout1 and Layout2. These two preset layouts do not include anything, so if you want to use them, take the following steps:
  - Rename the file to the desired name.
  - Create a page setup and link the layout to it, or you can simply link an existing page setup.
  - Insert a title block.
  - Insert Viewports, scale them, and lock them.

9.3 CREATING A NEW LAYOUT FROM SCRATCH

- This method allows you to create a new layout from scratch, as follows:
  - Right-click on any existing layout name (the tab at the lower-left corner of the screen and above the Command window) and a shortcut menu will appear; select the **New Layout** option:

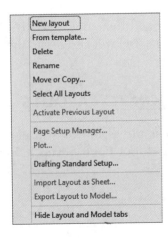

- A new layout will be added with a temporary name. You should rename it by right-clicking and selecting the **Rename** option (you can also click the temporary name, it will become editable, and you can input the new name):

- Link the new layout with a Page Setup. To do that, right-click the name of the layout, and then select **Page Setup Manager**:

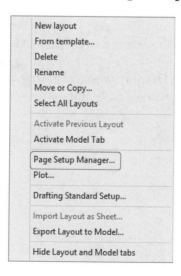

- The following dialog box will be shown:

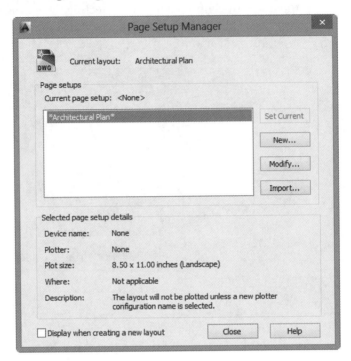

- At the top of the dialog box, you will see the name of the **Current layout**, at the bottom you will see the **Selected page setup details**, which is a summary of the current page setup. Finally, you will see a check box, **Display when creating a new layout**. This checkbox will force this dialog box to appear every time you go to a newly created layout.

- By default, AutoCAD will create a page setup with the same name of the layout. Click the **Modify** button to modify it, and you will see the following dialog box:

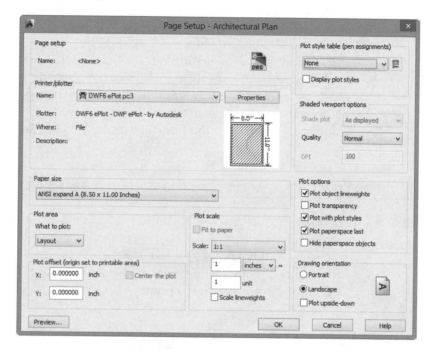

- Select the **Plotter** (this plotter should be connected and configured).
- Select the **Paper Size**.
- Specify **What to plot**. Always leave it as **Layout** (the other option is for Model space printing).
- Input **Plot Offset**. If you are printing from layout, leave the values as zeros.
- Input **Plot Scale**. If you are printing from layout then you will use viewports (this is our next topic), each viewport will have its own scale. So, you will set the layout plot scale to 1 = 1. Also specify if you want to **Scale lineweights**.
- Select **Plot style table (pen assignment)** (this topic will be discussed at the end of this chapter).
- **Shaded viewport options** are for 3D modeling in AutoCAD.

- Leave **Plot options** as the default values.
- Select **Drawing orientation**, whether **Portrait** or **Landscape**.
- The plotter will print from top-to-bottom; select the check box if you want it to print otherwise.
- Click **OK**. The Page Setup you create will be available for all layouts in the current drawing file.
- You will return to the first dialog box. Select the page setup and click **Set Current**. (You can also double click the name of the Page Setup.) Now the current layout is linked to page setup you selected.
- To modify the settings of an existing page setup click **Modify**. To import a saved page setup from an existing file click **Import**.

9.4 CREATING A NEW LAYOUT USING A TEMPLATE

- This procedure allows you to import a layout from a template file, including the page setup and any other contents like title block, viewports, text, etc. To do this, take the following steps:
  - Right-click on any existing layout you will see the following menu; select the **From template** option:

- The Open file dialog box will pop up so you can select the desired template file. Select the template, click **Open**, and you will see the following dialog box:

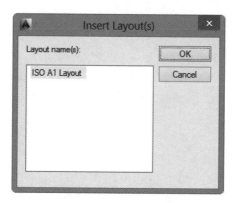

- Click one of the listed layouts and click **OK**. You will see the newly imported layout in your drawing.

9.5 CREATING LAYOUTS USING COPYING

- These two methods allow you to create copies of an existing layout. The first method is similar to a method used in Microsoft Excel. Take the following steps:
  - Select the desired layout.
  - Hold the [Ctrl] key on the keyboard and hold and drag the mouse to the new position of the newly copied layout:

  - Rename the new layout.

- Another way to copy layouts is to select the desired layout and then right-click, a shortcut menu will appear. Select the **Move or Copy** option:

- You will see the following dialog box:

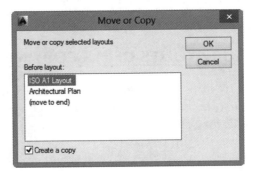

- A list of the current layouts is displayed. Select one of them and click the checkbox **Create a copy** on. A copy will be created; you can rename it and make the necessary changes.
- Using the same option you can move a layout from its current position to the left or to the right. Alternatively, you can move a layout without this command by clicking the layout name, holding it, and dragging it to the desired location.
- So far, you have:
  - Created a new layout
  - Created a page setup
  - Linked a page setup to the layout

- Accordingly, when you select the newly created layout, you will see something like the following:

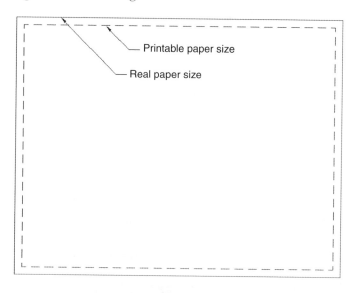

- The outer frame is the real paper size, and the inside frame (the dashed line) is the printable paper size, which is the paper size minus the printer's margins. Based on this layout, you will see exactly what will be printed and what will not, because anything outside the dashed line will not be printed. This proves that printing from layouts is WYSIWYG (What You See Is What You Get).
- You can reach some of the commands mentioned above if you go to the **Layout** tab and locate the **Layout** panel. You will see the following buttons:

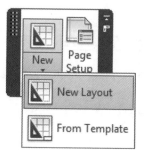

- In this panel, you can create a new layout from scratch or from a template. The second button allows you to access the Page Setup dialog box.

PRACTICE 9-1

Creating New Layouts

1. Start AutoCAD 2014.
2. Open **Practice 9-1.dwg**.
3. Select Layout 1 and delete the existing viewport.
4. Click then right-click Layout 1 and select Page Setup Manager.
5. Create a new Page Setup using the following information:
   a. Name = Architectural Plan A3
   b. Plotter = DWF6 ePlot.pc3
   c. Paper size = ISO A3 (420.00 × 297.00 MM)
   d. What to plot = Layout
   e. Plot scale = 1:1
   f. Drawing orientation = Landscape
   g. Click OK to end the creation process
6. Make sure you are selecting the newly created Page Setup and click Set Current to link this Page Setup to the current layout.
7. Rename the layout Architectural Plan (A3 Size).
8. Make layer Title Block the current layer.
9. Insert the file called A3 Size Title Block.dwg to be your title block, using the insertion point 0,0.
10. Create a copy of the newly created layout and name it Mechanical.
11. Delete Layout 2.
12. Right-click any existing layout and select the From template option.
13. Import the ISO A1 Layout from *Tutorial-mArch.dwt*.
14. Move the ISO A1 Layout to be the first layout after the Model tab.
15. Save and close the file.

9.6 CREATING VIEWPORTS

- When you visit a layout for the first time (just after creating it) you will see a single viewport in the center. A viewport is a window of any shape containing the view of your Model space, initially scaled to the size of the window.

- Viewports inserted in layouts can be tiled, will be scaled, and can be printed.
- You can add viewports to layouts by:
  - Adding a single rectangular viewport
  - Adding multiple rectangular viewports
  - Adding a single polygonal viewport
  - Converting an object to be a viewport
  - Clipping an existing viewport

9.6.1 Adding Single Rectangular Viewports

- This command allows you to add single rectangular viewports in a layout. You should specify two opposite corners to specify the area of the viewport. To issue this command, go to the **Layout** tab, locate the **Layout Viewports** panel, and then select the **Rectangular** button:

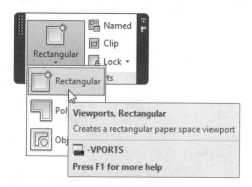

- The following prompts will be shown:

```
Specify corner of viewport or [ON/OFF/Fit/Shadeplot /
Lock/Object/Polygonal/Restore/LAyer/2/3/4] <Fit>:
Specify opposite corner:
```

- Specify the two opposite corners to create a single rectangular viewport.

■ This is what you will get:

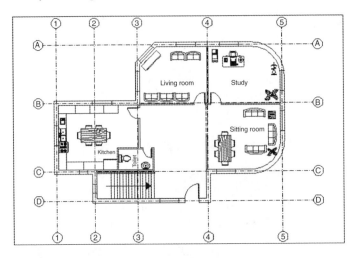

9.6.2 Adding Multiple Rectangular Viewports

■ This command allows you to add multiple rectangular viewports to a layout. You should specify two opposite corners for the area of the viewports. To issue this command, at the Command window type **+vports**, then press [Enter]. Press [Enter] again to accept the default value for the next prompt, and you will see the following dialog box:

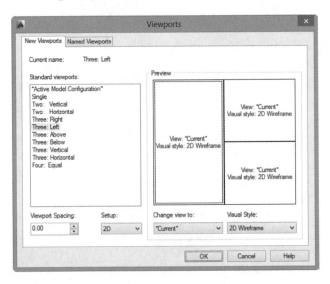

- Select the desired display from the list. By default the **Viewport Spacing** value = 0 (zero), which means the viewports will be tiled. If you want them separated, input a value greater than 0 (zero). Click **OK**, and you will see the following prompts:

```
Specify first corner or [Fit] <Fit>:
Specify opposite corner:
```

- See the following illustration:

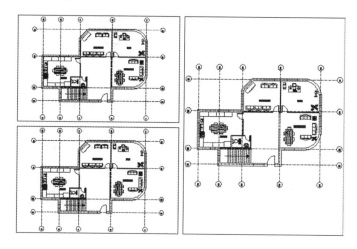

9.6.3 Adding a Polygonal Viewport

- This command allows you to add a polygonal viewport consisting of both straight lines and arcs. To start this command, go to the **Layout** tab, locate the **Layout Viewports** panel, and then select the **Polygonal** button:

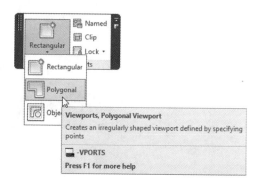

- You will see the following prompts:

```
Specify start point:
Specify next point or [Arc/Length/Undo]:
Specify next point or [Arc/Close/Length/Undo]:
```

- These prompts look like the Polyline command prompts.
- See the following illustration:

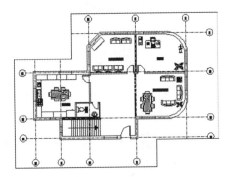

9.6.4 Creating Viewports by Converting Existing Objects

- This command allows you to convert any existing object (should be a single object, such as a polyline or circle; lines and arcs are not allowed) to the viewport. To start this command, go to the **Layout** tab, locate the **Layout Viewports** panel, and then select the **From Object** button:

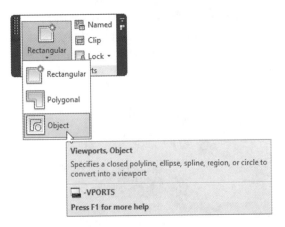

- You will see the following prompt:

```
Select object to clip viewport:
```

- See the illustration below:

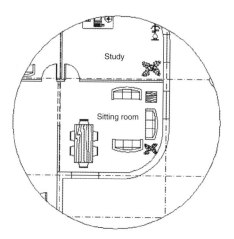

9.6.5 Creating Viewports by Clipping Existing Viewports

- This command allows you to clip an existing viewport and create a new shape. To start this command, go to the **Layout** tab, locate the **Layout Viewports** panel, and then select the **Clip** button:

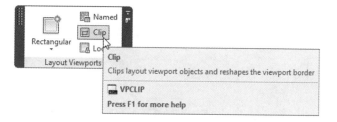

- The following prompts will appear:

```
Select viewport to clip:
Select clipping object or [Polygonal] <Polygonal>:
Specify start point:
Specify next point or [Arc/Length/Undo]:
Specify next point or [Arc/Close/Length/Undo]:
```

- First, select an existing viewport. You can select a polyline drawn previously, or you can draw any irregular shape using the **Polygonal** option (which is identical to the Polygonal viewport command discussed above). See the illustration below:

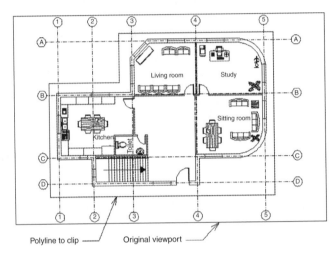

9.6.6 Working with Viewports

- There are two ways to work with viewports:
  - Outside the viewport, which means you will work with it like any other object in your drawing. You can select the viewport from its frame, and erase, copy, move, scale, stretch, rotate, etc.:

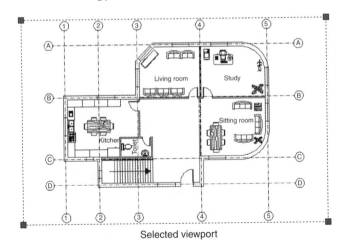

Selected viewport

- The second way is from inside the viewport, which you can achieve by double-clicking inside the viewport. This mode allows you to zoom, pan, scale, etc., the objects inside the viewport. To return to the first mode, double-click outside the viewport:

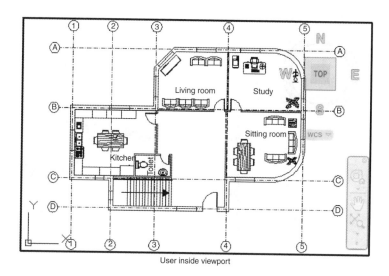

User inside viewport

9.7 SCALING AND MAXIMIZING VIEWPORTS

- By default, when you insert a viewport AutoCAD will zoom the whole drawing into the area you specified as the viewport size. This viewport is *not-to-scale*, and you should set the scale relative to the Model space units. Take the following steps:
  - Double-click inside the desired viewport (or you can select only the viewport's frame)
  - Look at the right side of the status bar, and you will see Viewport Scale list as shown below:

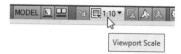

- Click the list that contains all scales, something like the following:

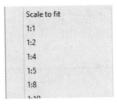

- Select the desired scale for your viewport. You can also select the **Custom** option, and the following dialog box will be displayed:

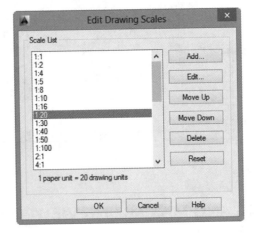

- To add a new scale, select the **Add** button, and the following dialog box will be displayed:

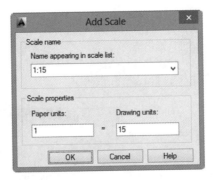

- Input the desired scale, and then click **OK** twice.
- After the scaling process is finished, either the scale will be perfect for the area of the viewport or it will be either too big or too small. The solution for such a problem is to change the size of the viewport or change the scale.
- After setting the scale, you can use the Pan command but not the Zoom command because this will ruin the scale you set. To avoid this problem, you can lock the display of the viewport by selecting the viewport(s) and then clicking the golden open lock in the status bar (you have to be inside the viewport or have the border of the viewport selected). The golden lock will change to a blue lock, and the viewports will be locked:

- The Maximize function allows you to maximize the viewport to fit the size of the screen temporarily. This will give you the space to edit objects as you wish without leaving the layout and going to Model space. Go to the status bar and click the **Maximize Viewport** button as shown:

- The same button will restore the original size of the viewport. Or, you can go to the **Layout** tab, the **Layout Viewports** panel, and use the following two buttons:

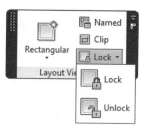

9.8 FREEZING LAYERS IN A VIEWPORT

- The Freezing function will freeze layer(s) in Model space, and all viewports in all layouts. If you want to freeze a layer in a certain viewport, take the following steps:
  - Double-click the desired viewport.
  - Go to the **Home** tab, locate the **Layers** panel, select the layer list, and click the icon **Freeze or thaw in current viewport** for the desired layer. See the illustration below:

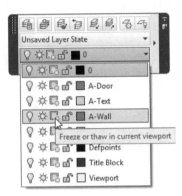

- You also have the ability to freeze/thaw the layer in all viewports except the current viewport. To do this start the Layer Properties Manager, select the desired layer(s), right-click, select **VP Freeze layer**, and then the **In All Viewports Except Current** option, as shown below:

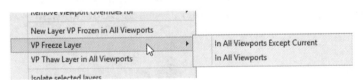

9.9 LAYER OVERRIDE IN A VIEWPORT

- In all viewports, a layer will be displayed with the same color, linetype, lineweight, and plot style, but you can change these settings in one viewport. This is called layer override. Take the following steps:
  - Double-click inside the desired viewport.
  - Issue the Layer Properties Manager command.

- Under VP Color, VP Linetype, VP Lineweight, or VP Plot Style make the desired changes.
- These changes will only affect the current viewport.
  - See the following illustration:

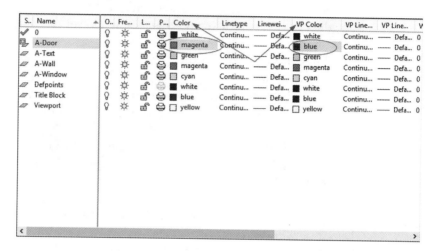

- In the above example, layer **A-Door** has a color (applies in Model space and all other viewports) of Magenta, and an override color of blue in the current viewport.
- Also notice that the layer **A-Door** row is shaded with a different color.

PRACTICE 9-2

Creating and Controlling Viewports

1. Start AutoCAD 2014.
2. Open **Practice 9-2.dwg**.
3. Make layer Viewport the current layer.
4. Switch to the D-Size Architectural Plan layout.
5. Insert a single viewport to fill the entire area.
6. Set the scale to ½" = 1' and lock the viewport.
7. Switch to the D-Size Arch Details layout.
8. Insert viewports using Three: Right, with Viewport Spacing = 0.25, and select an area to fill the paper size.

9. Select the borders of the big viewport at the right and scale it to be ¼" = 1'.
10. Using the grips, shrink the area of the viewport to something suitable for the scale chosen.
11. Select the two viewports at the left and set the scale to ¾" = 1'.
12. For the top viewport, using the Pan command set the view to show the Kitchen and Toilet.
13. For the bottom viewport, using the Pan command set the view to show the Living Room.
14. Select the three viewports and lock them.
15. Switch to the ANSI B Size Architectural Plan layout.
16. Draw a big circle in the center of the paper using the Create from Object command, and convert the circle to a viewport.
17. Double-click inside the new viewport, set the scale to ½" = 1', and then pan to show the Study.
18. In the current viewport, thaw Centerline and Centerline-TAGS.
19. Switch to the other layouts to make sure that these two layers were frozen only in this viewport.
20. Change the color of layer A-Walls in this viewport to red.
21. Save and close the file.

9.10 PLOT COMMAND

- This is our final step. This command sends whatever we set in the layout to the specified plotter. To issue this command, go to the **Output** tab, locate the **Plot** panel, and then select the **Plot** icon:

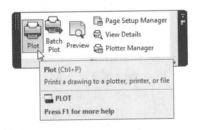

- You will see the following dialog box:

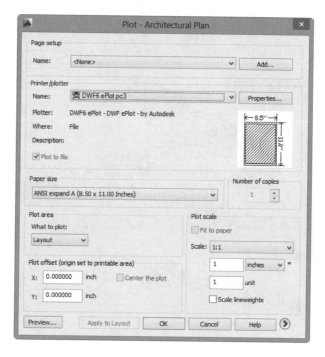

- This dialog box is identical to the Page setup settings. If you modify any of these settings, AutoCAD will separate the page setup from the current layout. To send the layout to the plotter, click **OK**.
- To save the settings of this dialog box with the layout, select the **Apply to Layout** button.
- Click the **Preview** button to preview the final printed drawing on the screen before it is printed.
- You can also preview your drawing from outside this dialog box by going to the **Output** tab, locating the **Plot** panel, and then selecting the **Preview** button:

NOTES:

CHAPTER REVIEW

1. Which of the following is true about creating a new layout?
    a. You can bring a layout from an existing template.
    b. You can create a new layout by right-clicking the name of an existing layout.
    c. You can bring in a layout using Design Center.
    d. All of the above.
2. Objects to be converted to a viewport should be _____ objects such as a circle or a polyline.
3. Using the Pan command after setting the scale of a viewport will ruin the scaling process.
    a. True
    b. Flase
3. You can freeze a layer in a viewport. You can change the color of a layer in a viewport.
    a. The first statement is correct, but the second is incorrect.
    b. Both statements are correct.
    c. Both statements are incorrect.
    d. The first statement is incorrect, but the second is correct.
4. _____ involves specifying plotter, page size, and orientation.

CHAPTER REVIEW ANSWERS

1. d
3. b
5. Page Setup

Chapter 10 PROJECTS

10.1 HOW TO PREPARE YOUR DRAWING FOR A NEW PROJECT

- In order to prepare your drawing for a new project, take the following steps:
  - Start a new drawing based on *acad.dwt* or *acadiso.dwt*.
  - Set up the Drawing Units. From the Application Menu select **Drawing Utilities/Units**:

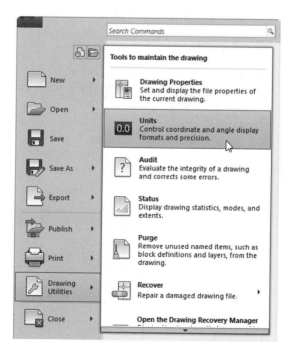

- You will see the following dialog box:

- Choose the desired **Length Type**. Select one of the following:
- Architectural (example: 2'-4 8/16"), Decimal (example: 25.5697), Engineering (example: 3'-5.6688"), Fractional (example: 16 3/16), or Scientific (example: 8.9643E+03).
- Choose the **Angle Type**. Select one of the following: Decimal Degrees (example: 36.7), Deg/Min/Sec (example: 47d25'31"), Grads (example: 60.7g), Radians (example: 0.6r), or Surveyor's Units (example: N 51d25'31" E).
- Set up the **Precision**, for both length and angle units. For example, choose the number of decimal for decimal units, between zero and eight decimal places.
- The default AutoCAD setting for angles is Counter-Clockwise, but you can switch it to **Clockwise.**
- Under **Insertion scale**, specify **Units to scale inserted content** (this was discussed in Chapter 6).
- Select the **Direction** button, and you will see the following dialog box:

- Change **East** to be 0 (zero) angle, and the other angles will change as well (we recommend this setting to be left as is).
- This step will end the Units command.
- Set up the **Drawing Limits**. The drawing limits contain your working area. You will specify the limits using two opposite corners; the lower-left corner and the upper-right corner. To set up the drawing limits correctly, answer two questions: what is the longest dimension in my drawing in both X and Y? And what is my AutoCAD unit (meter, centimeter, inch, or foot)?
- If the menu bar is showing, choose **Format/Drawing Limits**, or you can type **limits** at the Command window. You will see the following prompts:

```
Specify lower left corner or [ON/OFF] <0,0>:
Specify upper right corner <12,9>:
```

- Specify the lower-left corner and upper-right corner by typing or by clicking. To keep yourself from using any area outside these limits, AutoCAD allows you to turn it on and off.
- Create layers.
- Start drafting.

10.2 ARCHITECTURAL PROJECT (IMPERIAL)

 Take the following steps:

1. From the designated folder, open *Ground Floor_Starter.dwg*.
2. Switch off the grid.
3. Set the units to the following:
   a. Architectural
   b. Precision 0'- ½"
   c. Units to scale inserted contents = inch
4. Set up the drawing limits:
   a. Lower-left corner = 0,0
   b. Upper-right corner = 50', 50'
5. Double-click the mouse wheel in order to see the new limits.

6. Create the following layers:

Layer Name	Layer Color
A-Door	Blue
A-Wall	White
A-Window	White
Dimension	Red
Furniture	White
Staircase	Magenta
Text	Green
Title Block	White
Viewport	9
Hatch	8

7. Save as your file in the Chapter 10\Imperial folder and name it *Ground Floor.dwg*.
8. Make layer A-Wall current.
9. Draw the following architectural plan and the partitions inside using the following guidelines:
   a. Draw the outer shape using polyline.
   b. Offset it to the inside using 6" as distance.
   c. Explode the two polylines.
   d. Use the outer wall to draw the inner walls using all the commands you have learned. The inner wall is 4".
10. This is the architectural plan:

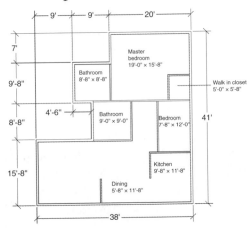

11. Create a 36" door opening as follows (you can always take a 4" clearance from the wall). The main entrance, master bedroom, and walk-in closet door are in the middle of the wall:

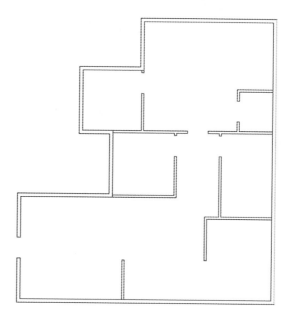

12. Make layer 0 (zero) current.
13. Create the following door blocks using these names (the base point is the lower-left point of the jamb):

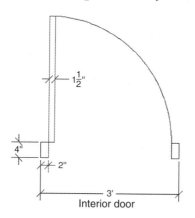

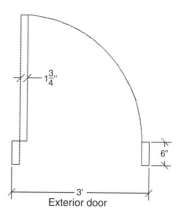

14. Create the following door block using this name (the base point is the lower-left point of the jamb):

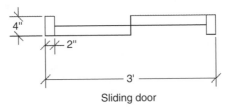

Sliding door

15. Create the following window blocks using these names (the base point is the lower-left point of the jamb):

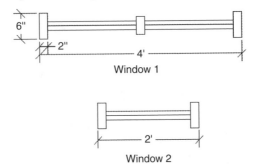

Window 1

Window 2

16. Insert the doors and windows in their respective layers to get the following result:

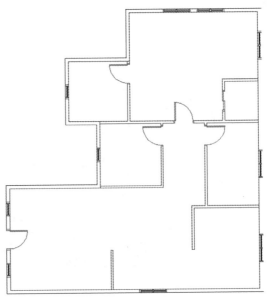

17. Make layer Furniture current.
18. Using the Insert command insert the following blocks:

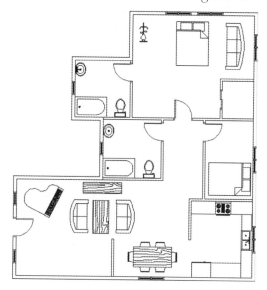

19. Make layer Hatch current.
20. Using Solid hatching hatch both the outside and inside wall.
21. Using ANSI37 and scale = 100 hatch the kitchen (hint: draw a line to separate the kitchen from the adjacent room).
22. Using the User-defined hatch (click the Double checkbox on) and scale = 20 hatch both bathrooms.
23. You should have something like the following:

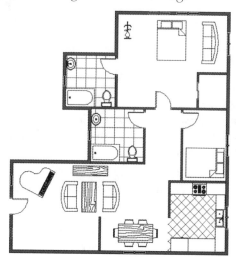

24. Make layer Text current, and freeze layer Hatch.
25. Add text using Multiline text to add the room titles just like the following, making sure Justify is Middle Center:

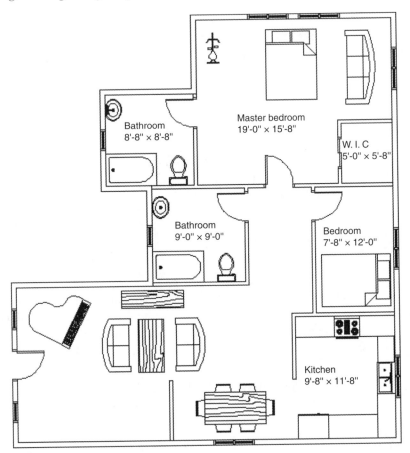

26. Thaw layer Hatch.
27. Select the hatch of one of the two bathrooms. When the context tab appears, locate the **Boundaries** panel and click the **Select** button, then choose the text, press [Enter], and then press [Esc]. Do the same for the other bathroom and the kitchen.
28. Make Outside Walls the current dimension style.
29. Make layer Dimension current.

30. Insert the dimensions as shown below (use the Continue command whenever possible):

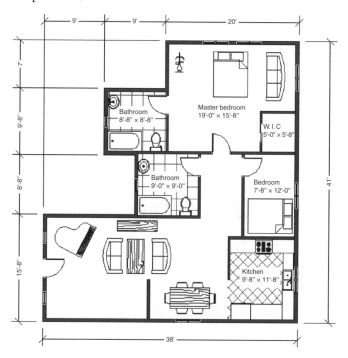

31. Go to Layout1 and rename it Full Plan.
32. Using the Page Setup Manager modify the existing page setup as follows:
    a. Printer = DWF6 ePlot.pc3
    b. Paper = ANSI B (17x11 inch)
    c. Drawing orientation = Landscape
33. Erase the existing viewport.
34. Make layer Title Block current.
35. Insert the file *ANSI B Landscape Title Block.dwg* in the layout using 0,0,0 as the insertion point.
36. Create a copy of the layout and name it Details.
37. Erase Layout 2.
38. Go to the layout 1 named Full Plan.
39. Make layer Viewport current.
40. Insert a single viewport to fill the space, and set the viewport scale to 3/16" = 1', then lock the viewport.

41. Go to Details layout.
42. Create a single viewport to occupy half of the space of the paper.
43. Set the scale to ½" = 1" and lock the viewport.
44. Pan to the entrance making sure to show the dimension, the two windows, and the door.
45. You are still in the Details layout. Create another viewport to occupy half of the remaining area of the paper.
46. Set the scale to ¼" = 1', then lock the viewport.
47. Double-click inside the viewport and pan to the two windows of the master bedroom.
48. Save and close the file.

10.3 ARCHITECTURAL PROJECT (METRIC)

Take the following steps:

1. From the designated folder, open *Ground Floor_Starter.dwg*.
2. Switch off the grid.
3. Set the units to the following:
   a. Decimal
   b. Precision = 0
   c. Units to scale inserted contents = Millimeters
4. Set up the drawing limits:
   a. Lower-left corner = 0,0
   b. Upper-right corner = 15000,15000
5. Double-click the mouse wheel in order to see the new limits.
6. Create the following layers:

Layer Name	Layer Color
A-Door	Blue
A-Wall	White
A-Window	White
Dimension	Red
Furniture	White

(Continued)

Layer Name	Layer Color
Staircase	Magenta
Text	Green
Title Block	White
Viewport	9
Hatch	8

7. Save as your file in Chapter 10\Metric folder and name it *Ground Floor.dwg*.
8. Make layer A-Wall current.
9. Draw the following architectural plan and the partitions inside using these guidelines:
   a. Draw the outer shape using a polyline.
   b. Offset it to the inside using 150 as distance.
   c. Explode the two polylines.
   d. Use the outer wall to draw the inner walls using all the commands you have learned. The inner wall is 100.
10. This is the architectural plan:

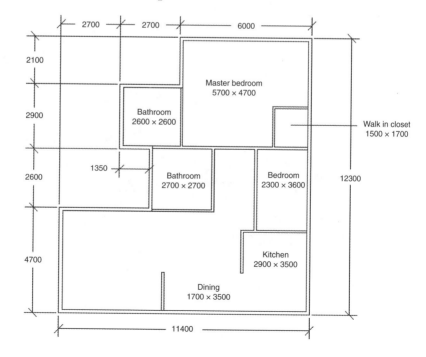

11. Create a 900 door opening as follows (you can always take a 100 clearance from the wall). The main entrance, master bedroom, and walk-in closet door are in the middle of the wall:

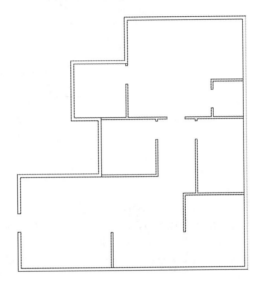

12. Make layer 0 (zero) current.
13. Create the following door blocks using these name (the base point is the lower-left point of the jamb).

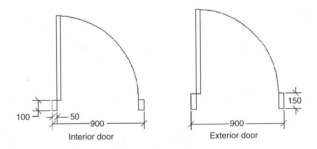

Interior door   Exterior door

14. Create the following door block using this name (the base point is the lower-left point of the jamb):

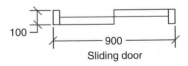

Sliding door

15. Create the following window blocks using these names (the base point is the lower-left point of the jamb):

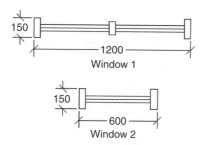

16. Insert the doors and windows in their respective layers to get the following result:

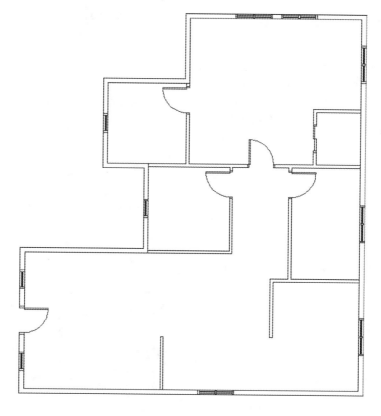

17. Make layer Furniture current.

18. Using the Insert command insert the following blocks:

19. Make layer Hatch current.
20. Using Solid hatching hatch both the outside and inside wall.
21. Using ANSI37 and scale = 2000 hatch the kitchen (hint: draw a line to separate the kitchen from the adjacent room).
22. Using the User-defined hatch (click the Double checkbox on) and scale = 500 hatch both bathrooms.
23. You should have something like the following:

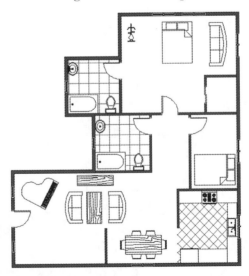

24. Make layer Text current, and freeze layer Hatch.
25. Make Room Titles the current text style.
26. Add text using Multiline text to add the room titles just like the following, making sure Justify is Middle Center:

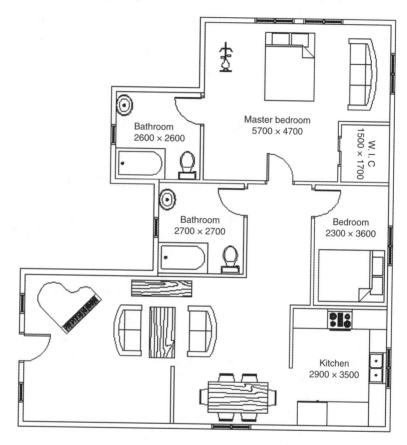

Bathroom
2600 × 2600

Master bedroom
5700 × 4700

W. I. C
1500 × 1700

Bathroom
2700 × 2700

Bedroom
2300 × 3600

Kitchen
2900 × 3500

27. Thaw layer Hatch.
28. Select the hatch of one of the two bathrooms. When the context tab appears, locate the **Boundaries** panel, click the **Select** button, and then choose the text, press [Enter], and then press [Esc]. Do the same for other the bathroom and the kitchen.
29. Make Outside Walls the current dimension style.
30. Make layer Dimension current.

31. Insert the dimensions as shown below (use the Continue command whenever possible):

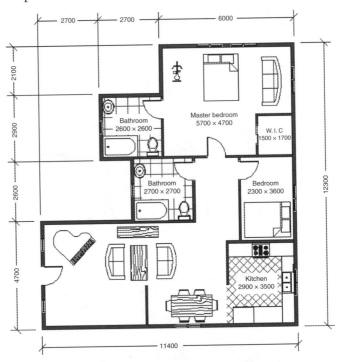

32. Go to Layout1 and name it Full Plan.
33. Using the Page Setup Manager modify the existing page setup as follows:
    a. Printer = DWF6 ePlot.pc3
    b. Paper = ISO A3 (420 × 297 MM)
    c. Drawing orientation = Landscape
34. Erase the existing viewport.
35. Make layer Title Block current.
36. Insert the file *ISO A3 Landscape Title Block.dwg* in the layout, using 0,0,0 as the insertion point.
37. Create a copy of the layout and name it Details.
38. Erase Layout 2.
39. Go to layout named Full Plan.
40. Make layer Viewport current.
41. Insert a single viewport to fill the space, and set the viewport scale to 1:100, then lock the viewport.

42. Go to Details layout.
43. Create a single viewport to occupy half of the space of the paper.
44. Set the scale to 1:20 and lock the viewport.
45. Pan to the entrance making sure to show the dimension, the two windows, and the door.
46. Make layer Viewport current.
47. You are still in Details layout. Create another viewport to occupy half of the remaining area of the paper.
48. Set the scale to 1:40, then lock the viewport.
49. Double-click inside the viewport, and pan to the two windows of the master bedroom.
50. Save and close the file.

10.4 MECHANICAL PROJECT – I (METRIC)

Take the following steps:

1. From the designated folder, open *Mechanical-1_Starter.dwg*.
2. Switch off the grid.
3. Set the units to the following:
   a. Decimal
   b. Precision 0
   c. Units to scale inserted contents = Millimeters
4. Set up the drawing limits:
   a. Lower-left corner = 0,0
   b. Upper-right corner = 350,250
5. Double-click the mouse wheel in order to see the new limits.
6. Create the following layers:

Layer Name	Layer Color	Layer Linetype
Centerline	Green	Center × 2
Dimension	Red	Continuous
Hatch	8	Continuous
Hidden	Cyan	Hidden × 2
Part	White	Continuous

(Continued)

Layer Name	Layer Color	Layer Linetype
Text	Magenta	Continuous
Title Block	White	Continuous
Viewport	9	Continuous

7. Save as your file in Chapter 10\Metric folder and name it *Mechanical–1.dwg*.
8. Make layer Part current.
9. Draw the following plan, section, and elevation of the mechanical part using the following guidelines:
   a. All lines of the shape in layer Part
   b. All centerlines in layer Centerline
   c. All hidden lines in layer Hidden
   d. Change the **Linetype scale** using the **Properties** palette for both centerlines and hidden lines to 5, except for the two holes at the right and left, which should be 2.
10. Draw the shape without dimensioning for now:

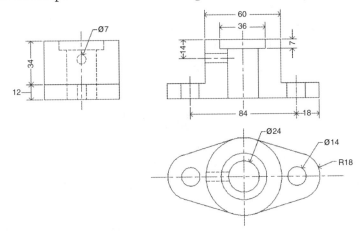

11. Make layer Hatch current.
12. Using ANSI31 with scale = 20 hatch the shape as shown below:

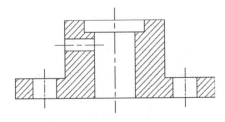

13. Make Part Dim the current dimension style.
14. Make layer Dimension current. Insert the dimensions as shown below:

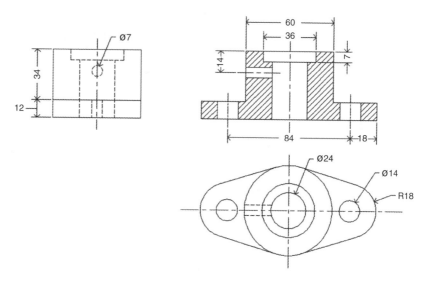

15. Go to Layout1 and name it Plan.
16. Using the Page Setup Manager modify the existing page setup as follows:
    a. Printer = DWF6 ePlot.pc3
    b. Paper = ISO A3 (420 × 297 MM)
    c. Drawing orientation = Landscape
    d. Make sure Scale = 1:1
17. Erase the existing viewport.
18. Make layer Title Block current.
19. Insert the file *ISO A3 Landscape Title Block.dwg* in the layout using 0,0,0 as the insertion point.
20. Make two copies of the Plan layout and name them Section and Elevation.
21. Delete Layout 2.
22. Make layer Viewport current.
23. Insert a single viewport with the following settings:
    a. Set the scale to 2:1
    b. Lock the viewport
    c. Pan to the shape as shown below
    d. You will notice that the hidden lines and centerlines looks like continuous lines. To solve this problem, at the command window type

the **psltscale** command and set this variable to 0. Then type the
regenall command to regenerate all viewports, and you will get the
following result:

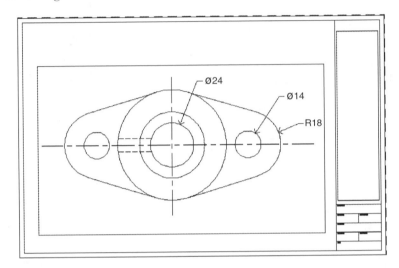

24. Repeat the same procedure to create a section viewport in the Section
layout and elevation viewport in the Elevation layout.
25. Save and close the file.

10.5 MECHANICAL PROJECT I (IMPERIAL)

 Take the following steps:

1. From the designated folder, open *Mechanical-1_Starter.dwg*.
2. Switch off the grid.
3. Set the units to the following:
   a. Fractional
   b. Precision 0 – 1/16
   c. Units to scale inserted contents = Inches
4. Set up the drawing limits:
   a. Lower left corner = 0,0
   b. Upper right corner = 18",9"
5. Double-click the mouse wheel in order to see the new limits.

6. Create the following layers:

Layer Name	Layer Color	Layer Linetype
Centerline	Green	Center2
Dimension	Red	Continuous
Hatch	8	Continuous
Hidden	Cyan	Hidden2
Part	White	Continuous
Text	Magenta	Continuous
Title Block	White	Continuous
Viewport	9	Continuous

7. Save as your file in the Chapter 10\Imperial folder and name it *Mechanical–1.dwg*.
8. Make layer Part current.
9. Draw the following plan, section, and elevation of the mechanical part using the following guidelines:
   a. All lines of the shape in layer Part
   b. All centerlines in layer Centerline
   c. All hidden lines in layer Hidden
   d. Change the **Linetype scale** using the **Properties** palette to 0.5 for any line you like.
10. Draw the shape without dimensions for now:

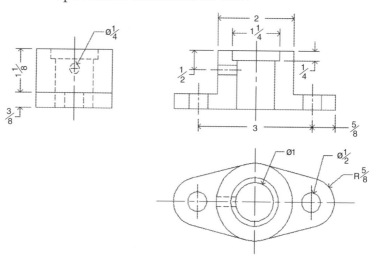

11. Make layer Hatch current.
12. Using ANSI31 with scale = 1 hatch the shape as shown below:

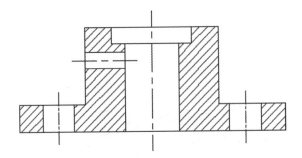

13. Make Part Dim the current dimension style.
14. Make layer Dimension current.
15. Insert the dimensions as shown below:

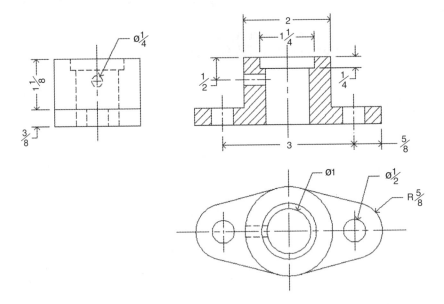

16. Go to Layout1 and name it Details.

17. Using Page Setup Manager modify the existing page setup as follows:
    a. Printer = DWF6 ePlot.pc3
    b. Paper = ANSI B (17x11 in)
    c. Drawing orientation = Landscape
    d. Make sure Scale = 1:1
18. Erase the existing viewport.
19. Make layer Title Block current.
20. Insert the file *ANSI B Landscape Title Block.dwg* in the layout using 0,0,0 as the insertion point.
21. Delete Layout 2.
22. Make layer Viewports current.
23. Insert three single viewports and do the following:
    a. Set the scale for the three viewports to 1' = 1'
    b. Lock the viewports
    c. Pan to the shape as shown below
    d. If you notice that hidden lines and centerlines look like continuous lines, at the Command window type the **psltscale** command and set this variable to 0. Then type the **regenall** command to regenerate all viewports, and you will get the following result:

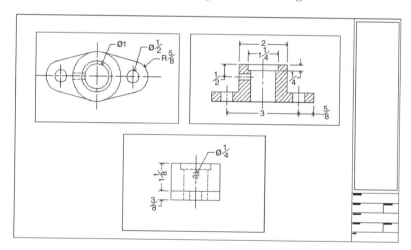

24. Save and close the file.

10.6 MECHANICAL PROJECT – II (METRIC)

Using the same methodology we used in Mechanical Project – I (Metric) draw the following project using *Mechanical-1_Starter.dwg*:

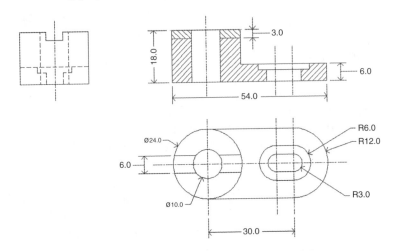

10.7 MECHANICAL PROJECT – II (IMPERIAL)

Using the same methodology we used in Mechanical Project – I (Imperial) draw the following project using *Mechanical-1_Starter.dwg*:

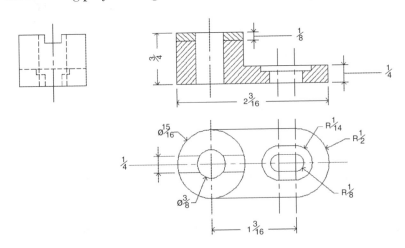

NOTES:

Chapter **11** ## MORE ON 2D OBJECTS

In This Chapter

◇ Polyline command and other drafting and editing commands
◇ How to use both Constructions Line and Ray commands
◇ How to use the Point command with different styles with the Divide and Measure commands
◇ Using the Spline and Ellipse commands
◇ Using the Boundary and Region commands with Boolean operations

11.1 INTRODUCTION

■ This chapter will discuss the 2D objects not discussed in Chapter 2. We will start with the Polyline command, since the other 2D objects are based on this command, and we will also cover special features of polylines in some of the editing commands. Next, we will discuss other 2D commands such as Spline and Ellipse. These two commands have unique features that allow you to draw exact 2D curves. Then we will delve into other commands such as Point (Divide and Measure), Revision Cloud, and Wipeout. Finally, we will end the chapter by discussing Boundary and Region commands.

11.2 DRAWING LINES AND ARCS USING THE POLYLINE COMMAND

■ The Polyline command allows you to do all or any of the following:
  • Draw both line segments and arc segments

- Draw single objects in the same command rather than drawing segments of lines and arcs, such as with the Line and Arc commands
- Draw lines and arcs with starting and ending widths
■ To use the command, go to the **Home** tab, locate the **Draw** panel, and then select the **Polyline** button:

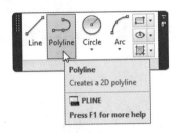

■ The following prompts will appear:

```
Specify start point:
Current line-width is 1.0000
Specify next point or [Arc/Halfwidth/
Length/Undo/ Width]:
```

■ AutoCAD asks you to specify the first point, and when you do, AutoCAD will give you the current line-width. If it works for you, continue specifying points using the same method we learned in the Line command; if not, change the width as the first step, type the letter **W**, or right-click and select the **Width** option, and you will see the following prompts:

```
Specify starting width <1.0000>:
Specify ending width <1.0000>:
```

■ Specify the starting width, press [Enter], and then specify the ending width. The next time you use the same file, AutoCAD will report these values to you when you issue the Polyline command. Halfwidth is the same, but instead of specifying the full width, you specify halfwidth.
■ The **Undo** and **Close** options are identical to the ones in the Line command.
■ Length allows you to specify the length of the line using the angle of the last segment.

- Arc allows you to draw an arc attached to the line segment. You will see the following prompt:

```
Specify endpoint of arc or [Angle/CEnter/CLose/
Direction/ Halfwidth/Line/Radius/Second
pt/Undo/Width]:
```

- The arc will be attached to the last segment of the line, or will be the first object in the Polyline command. Using either method, the first point of the arc is already known, so we need two more pieces. AutoCAD will make an assumption (that you have the right to reject) and will assume that the angle of the last line segment will be considered the direction (tangent) of the arc. If you accept this assumption, you should specify the endpoint. If not, choose from the following to specify the second piece of information:
  - The Angle of the Arc
  - The Center point of the arc
  - Another Direction to the arc
  - The Radius of the arc
  - The Second point, which can be any point on the parameter of the arc
- Based on the information selected as the second point, AutoCAD asks you to supply the third piece of information.

11.3 CONVERTING POLYLINES TO LINES AND ARCS, AND VICE-VERSA

- This is a very essential technique that allows you to convert any polyline to lines and arcs, and convert lines and arcs to polylines.

11.3.1 Converting Polylines to Lines and Arcs

- The **Explode command** will explode a polyline into lines and arcs. To issue this command, go to the **Home** tab, locate the **Modify** panel, and then select the **Explode** button:

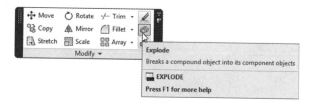

- AutoCAD will show the following prompt:

```
Select objects:
```

- Select the desired polylines and press [Enter] when done. The new shape will have lines and arcs.

11.3.2 Joining Lines and Arcs to Form a Polyline

- In this section, we will discuss an option called **Join**, within a command called **Edit Polyline**. To issue this command, go to the **Home** tab, locate the **Modify** panel, and then select the **Edit Polyline** button:

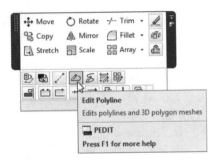

- You will see the following prompts:

```
Select polyline or [Multiple]:
Object selected is not a polyline
Do you want to turn it into one? <Y>
Enter an option [Close/Join/Width/Edit vertex/Fit/
Spline/Decurve/Ltype gen/Reverse/Undo]: J
```

- Begin the first command by selecting one of the lines or arcs you want to convert. AutoCAD will respond by telling you that the selected object is not a polyline! and gives you the option to convert this specific line or arc to a polyline. If you accept this, options will appear, and one of these options will be **Join**. Select the **Join** option, and then select the rest of the lines and arcs. At the end, press [Enter] twice. When you are finished, you will see that the objects were converted to a polyline.

PRACTICE 11-1

Drawing Polylines and Converting

1. Start AutoCAD 2014.
2. Open **Practice 11-1.dwg**.
3. Draw the following polyline using a start point of 18,5 and width = 0.1:

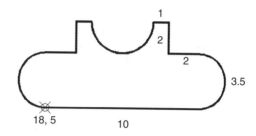

4. Then explode the polyline. As evidence, the width will disappear.
5. After exploding, the objects are lines and arcs.
6. Save and close.

11.4 DRAWING USING THE RECTANGLE COMMAND

- This command will draw a rectangle or square shape. The **Rectangle** command uses a polyline as an object. To issue this command, go to the **Home** tab, locate the **Draw** panel, and then select the **Rectangle** button:

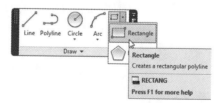

- You will see the following prompts:

```
Specify first corner point or [Chamfer/Elevation/
Fillet /Thickness/Width]:
Specify other corner point or [Area/Dimensions/
Rotation]:
```

- By default, you can draw a rectangle by specifying two opposite corners. The other options are discussed as follows.

11.4.1 Chamfer Option

- This option allows you to draw a rectangle with chamfered edges. You will see the following prompts:

```
Specify first chamfer distance for rectangles <0.00>:
Specify second chamfer distance for rectangles <0.2>:
```

- Specify the first and second distance, then the Rectangle command will continue with the normal prompts. See the following example:

11.4.2 Elevation Option

- This option is used for 3D only.

11.4.3 Fillet Option

- This option is identical to the Chamfer option, except you have to input Radius instead of Distance. You will see the following prompt:

```
Specify fillet radius for rectangles <0.0000>:
```

- Specify the fillet radius, then the Rectangle command will continue with normal prompts. See the following example:

11.4.4 Thickness Option

- This option is used for 3D only.

11.4.5 Width Option

- You can draw a rectangle with width using this option. You will see the following prompt:

```
Specify line width for rectangles <0.0000>:
```

- Specify the width value, then the Rectangle command will continue normally. See the following example:

11.4.6 Area Option

- This option allows you to specify the total area of the rectangle prior to specifying the second corner. You will see the following prompts:

```
Enter area of rectangle in current units <25.0000>:
Calculate rectangle dimensions based on
[Length/Width] <Length>:
Enter rectangle length <10.0000>:
```

- AutoCAD asks you to input the total area and then asks you to input either the length (in the X-axis) or width (in the Y-axis). AutoCAD will then draw a rectangle above and to the right of the first corner.

11.4.7 Dimensions Option

- This option allows you to draw a rectangle by specifying length (in the X-axis) and width (in the Y-axis). You will see the following prompts:

```
Specify length for rectangles <10.0000>:
Specify width for rectangles <10.0000>:
Specify other corner point or [Area/Dimensions/
Rotation]:
```

- AutoCAD asks you to input the length and the width. The final prompt asks you to input the position of the second point.

11.4.8 Rotation Option

- This option allows you to draw a rectangle with a rotation angle. You will see the following prompt:

  ```
  Specify rotation angle or [Choose points] <0>:
  ```

- Specify the rotation angle either by typing the value, or by specifying points.

11.5 DRAWING USING THE POLYGON COMMAND

- This command allows you to draw an equilateral polygon with 3 sides to 1024 sides. The Polygon command uses the polyline as an object. To issue this command, go to the **Home** tab, locate the **Draw** panel, and then select the **Polygon** button:

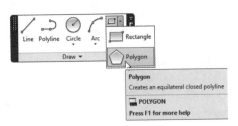

- You will see the following prompt:

  ```
  Enter number of sides <6>:
  ```

- Input the number of sides for your polygon. Next, you will see the following prompt:

  ```
  Specify center of polygon or [Edge]:
  ```

- AutoCAD offers two methods to draw a polygon. Either by using an imaginary circle, or by specifying the length and angle of one of the sides.

11.5.1 Using an Imaginary Circle

- This method depends on an imaginary circle. The polygon is either inscribed inside it, or circumscribed about it. The center of the circle and the polygon coincide, so the radius of the circle will determine the size of the polygon. The question is: When should you use each specific method? The answer to that question depends on the available information. If you know the distance between one of the edges and the polygon's center, then use the **Inscribed** option. But if you know the distance between the midpoint of one of the edges, and the center, then the **Circumscribed** option is the best solution. See the following illustration:

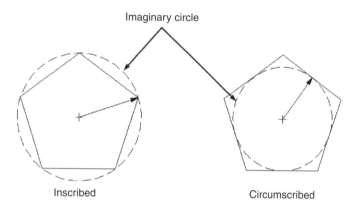

Inscribed   Circumscribed

- You will see the following prompts:

```
Enter an option [Inscribed in circle/
Circumscribed about circle] <I>:
Specify radius of circle:
```

11.5.2 Using the Length and Angle of One of the Edges

- If you don't know the center of the polygon, then you can't use the above method. But if you specify the length of one side, the other side's length will be known automatically. While you are specifying the two points as a length of one of the sides, you are also specifying the angle of this side, and accordingly the angles of the other sides will be defined. You will see the following prompts:

```
Enter number of sides <4>:
Specify center of polygon or [Edge]:
Specify first endpoint of edge:
Specify second endpoint of edge:
```

PRACTICE 11-2

Drawing Rectangles and Polygons

1. Start AutoCAD 2014.
1. Open **Practice 11-2.dwg**.
2. Use the Rectangle and Polygon commands to complete the practice as follows (using OSNAP = Node) to select the two points. You will need the commands in the upper Right including Explode, Extend, and Trim to get the correct results:

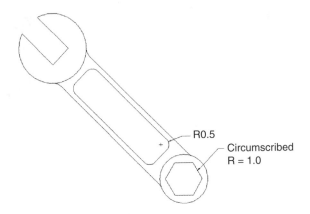

3. Freeze layer Points.
4. Save and close.

11.6 DRAWING USING THE DONUT COMMAND

- This command allows you to draw either a circle with width, or a filled circle. Donut uses a polyline as the object. To issue this command, go to the **Home** tab, locate the **Draw** panel, and then select the **Donut** button:

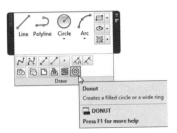

- The following prompts will appear:

```
Specify inside diameter of donut <0.5000>:
Specify outside diameter of donut <1.0000>:
Specify center of donut or <exit>:
```

- AutoCAD asks you to input the inside and outside diameter, then to specify the center of the donut. You can insert as many donuts as needed.

11.7 DRAWING USING THE REVISION CLOUD COMMAND

- This command allows you to draw a revision cloud using polyline arcs. To issue this command, go to the **Home** tab, locate the **Draw** panel, and then select the **Revision Cloud** button:

- You will see the following prompts:

```
Minimum arc length: 15 Maximum arc
length: 15 Style: Normal
Specify start point or [Arc length/
Object/Style] <Object>:
```

- You should input the minimum and maximum arc length, then specify the revision cloud style. If you select **Arc length**, you will see the following prompts:

```
Specify minimum length of arc <15>:
Specify maximum length of arc <30>:
```

- If you select **Style**, you will see the following prompt:

```
Select arc style [Normal/Calligraphy] <Normal>:
```

- To understand the difference between the two styles, see the following illustration:

- When these two settings are completed, simply specify the first point of the revision cloud, then move your mouse (don't click) in the desired direction, and AutoCAD will close the shape automatically and end the command once you get closer to the start point.
- Another way to draw a revision cloud is by converting a closed 2D object (circle, polyline, ellipse, etc.). You will see the following prompts:

```
Select object:
Reverse direction [Yes/No] <No>:
```

PRACTICE 11-3

 Drawing Donuts and Revision Clouds

1. Start AutoCAD 2014.
2. Open **Practice 11-3.dwg**.
3. Using the horizontal line and the vertical lines insert donuts with inside diameter = 0 and outside diameter = 0.25.
4. Freeze layer Scratch layer.
5. Make Redline the current layer.

6. Using the Revision Cloud command, change the min and max arc length to be 1, and draw and convert the circle to get the following result:

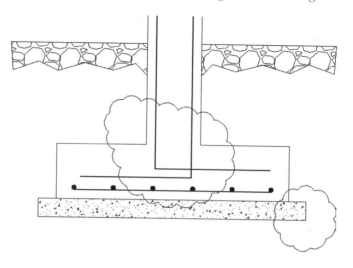

7. Save and close.

11.8 USING THE EDIT POLYLINE COMMAND

- This is a special editing command that can work only with polylines. It can do certain things that the normal modifying commands can't. We saw some of this command's power when we touched on the subject of converting lines and arcs to polylines. To issue this command, go to the **Home** tab, locate the **Modify** panel, and then select the **Edit Polyline** button:

- Another way is to double-click the polyline. You will see the following prompts:

```
Select polyline or [Multiple]:
Enter an option [Close/Join/Width/Edit vertex/Fit/
Spline/Decurve/Ltype gen/Reverse/Undo]:
```

- As you can see, AutoCAD asks you to select a single polyline to perform one of many editing options. AutoCAD can also work with **Multiple** polylines. We will first discuss the editing options for a single polyline, then we will cover options for multiple polylines.

11.8.1 Open and Close Options

- The **Open** option will be displayed if the selected polyline is closed, and vice-versa. There are not any prompts for those two options, as Auto-CAD remembers the last segment drawn and will erase it to create an opened polyline.

11.8.2 Join Option

- This option allows you to join lines and arcs to the first selected polyline. You will see the following prompt:

```
Select objects:
```

- You are invited to select objects to join them to the selected polyline.

11.8.3 Width Option

- This option allows you to specify a width for the selected polyline. You will see the following prompt:

```
Specify new width for all segments:
```

11.8.4 Edit Vertex Option

- This option allows you to select a vertex in the polyline, then perform an editing option on this vertex. Many AutoCAD users agree that this option is lengthy, tedious, and difficult. Instead, you can explode the polyline, perform all/any normal modifying commands, then join the lines and arcs in a single polyline. You will see the following prompt:

```
[Next/Previous/Break/Insert/Move/Regen/Straighten/
Tangent/Width/eXit] <N>:
```

11.8.5 Fit, Spline, and Decurve Options

- The **Fit** and **Spline** options will convert a straight-line segment polyline to a curved polyline using two different methods. See the following illustration:

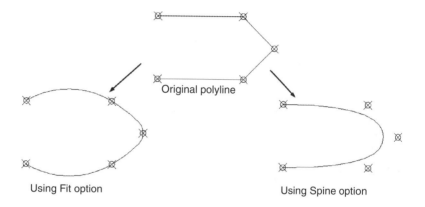

Original polyline

Using Fit option

Using Spine option

- As you can see from the above illustration, each command handles the process in a different way:
  - The **Fit** option uses the same vertices and connects them using a curve. So, this method is considered approximate. The **Spline** option uses the vertices as controlling points to draw the needed curve. This method will display a more accurate curve.
  - The **Decurve** option allows you to convert back from a curved shape polyline to a straight-line polyline.

11.8.6 Ltype gen Option

- This option allows you to convert polylines from straight lines to curved lines to retain the original linetype. See the following illustration:

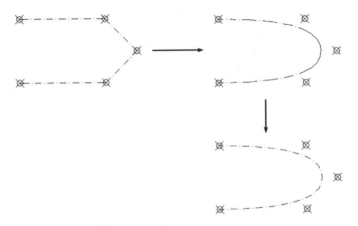

- You will see the following prompt:

```
Enter polyline linetype generation option
[ON/OFF] <Off>:
```

- In order to retain the linetype, input **ON** at this prompt.

11.8.7 Reverse Option

- This option allows you to reverse the order of vertices in a polyline. The results will be evident when using a special linetype such as the following:

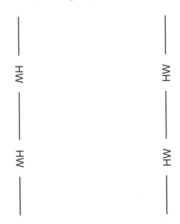

- There is a system variable called **PLINEREVRESEWIDTHS** that controls whether to reverse the polyline width or not. If the value is 0 (zero) the polyline will not be reversed, but if the value is 1, the polyline width will be reversed. See the following illustration:

PLINEREVRESEWIDTHS = 1

11.8.8 Multiple Option

- This command allows you to modify multiple polylines using a single modifying command. Another purpose for this command is to join polylines together. This is different than the Join option discussed above, which is for a single polyline that joins lines and arcs. This option is used to join polylines together in a single polyline. You will see the following prompts:

```
Select objects:
Select objects:
Enter an option [Close/Open/Join/Width/Fit/
Spline /Decurve/Ltype gen/Reverse/Undo]:
```

- These prompts are identical to editing a single polyline, except the **Edit Vertex** option. If you select the Join option, you will see the following prompt:

```
Join Type = Extend
Enter fuzz distance or [Jointype] <0.0000>:
```

- The first line gives you the current value for Join Type, which is Extend. To change it, use the **Jointype** option to see the following prompt:

```
Enter join type [Extend/Add/Both] <Extend>:
```

- There are three types of joining:
  - Extend: AutoCAD will extend the two ends to each other to join the multiple polylines.
  - Add: AutoCAD will add a line between the two ends.
  - Both: AutoCAD will join both methods.
- AutoCAD will also need to know the **Fuzz distance**, the maximum acceptable distance between the ends of the two polylines to join. Auto-CAD will not join polylines greater than this value. See the following illustration:

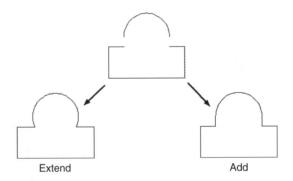

- One final note for the Polyline Edit command. Professional often cut steps while drafting and editing, and AutoCAD can make this process smoother. System variable **PEDITACCEPT** can be used to assume the answer to the following prompt while selecting the first object in the Polyline command to always be yes:

```
Object selected is not a polyline
Do you want to turn it into one? <Y>
```

- PEDITACCEPT has two values:
  - 0 (zero): AutoCAD asks the question and waits for you to confirm.
  - 1: AutoCAD assumes the answer to always be Yes.

PRACTICE 11-4

Polyline Edit Command

1. Start AutoCAD 2014.
2. Open **Practice 11-4.dwg**.
3. Notice the objects drawn. You will find that all objects are polylines except the outer contours, which are lines. Using the Polyline Edit command, convert the lines to polylines.
4. Using the Polyline Edit command, convert the straight-line segments to spline segments.
5. Using the Polyline Edit command, retain the dashed lines of the converted polylines.
6. Zoom to the two small buildings inside the contours, and you will find them to be polylines, but the arc is not reaching the line segments. Measure the void between the arc and the lines, and input a proper Fuzz distance. Then join the polylines, using Extend at the left shape and Add at the right shape.
7. You should get the following:

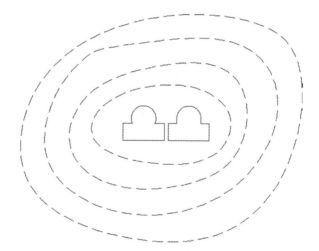

8. Save and close.

11.9 USING CONSTRUCTION LINES AND RAYS

- These two commands produce objects used as helping tools to draw accurate drawings; they are not useful on their own. Construction lines are objects extending beyond the screen in the two directions and can be drawn using different methods. Rays in AutoCAD are objects that have a known starting point, extending beyond the screen in the other direction. Both commands are discussed as follows.

11.9.1 Construction Lines Command

- This command helps you draw a construction line that extends beyond the screen in two directions. To issue this command, go to the **Home** tab, locate the **Draw** panel, and then select the **Construction Line** button:

- You will see the following prompt:

```
Specify a point or [Hor/Ver/Ang/Bisect/Offset]:
```

- There are six methods to specify the angle of the construction line:
  - The first method is the default method, which is to specify two points. Once you specify the first point, you will see the following prompt:

```
Specify through point:
```

  - The **Hor** and **Ver** options are used to draw horizontal or vertical construction lines. To complete the command, specify the through point. You will see the following prompt:

```
Specify through point:
```

- The **Ang** option is used to draw a construction line using an angle. You will see the following prompts:

```
Enter angle of xline (0) or [Reference]:
Specify through point:
```

- **The Bisect** option involves specifying three points, passing through the first point and bisecting the angle formed between the second and third points. You will see the following prompts:

```
Specify angle vertex point:
Specify angle start point:
Specify angle end point:
```

- The **Offset** option is used to produce a construction line parallel to an existing line. You will see the following prompts:

```
Specify offset distance or [Through] <Through>:
Select a line object:
Specify side to offset:
```

11.9.2 Rays Command

- This command allows you to draw a ray that has a starting point and an end that extends beyond the screen. To issue this command, go to the **Home** tab, locate the **Draw** panel, and select the **Ray** button:

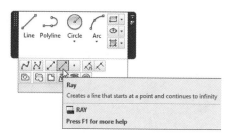

- You will see the following prompts:

```
Specify start point:
Specify through point:
```

- AutoCAD asks you to specify two points. The first point is the starting point, and the second will define the angle of the ray. You can define as many rays as you want, using the same starting point.

PRACTICE 11-5

Using Construction Lines and Rays

1. Start AutoCAD 2014.
2. Open **Practice 11-5.dwg**.
3. Make layer Construction current.
4. Insert vertical and horizontal construction lines using the center of the circle.
5. Insert two construction lines using the Offset option, using the two vertical lines with distance = 0.5, to the inside.
6. Using the Ray command set the starting point to be the center of the circle, and the second point to be with angle = 60 using Polar Tracking.
7. Repeat the same command, using angles 120, 240, and 300.
8. Draw a new circle with its center coinciding with the existing circle, using R = 2.
9. You should have the following:

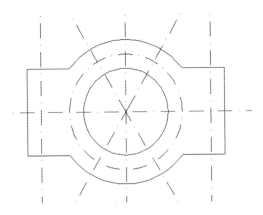

10. Make layer Circle current, and make sure Intersection in OSNAP is turned on.
11. Draw circles using the intersection of the four rays and the circle, along with the vertical construction line and the circle, using R = 0.2.

12. Draw two circles at the intersections of the horizontal and vertical construction lines.
13. Freeze layer Construction.
14. You should have the following:

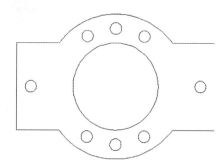

15. Save and close the file.

11.10 USING THE POINT STYLE AND POINT COMMANDS

- The Point Style command will set the shape of the point, and the Point command will insert a point in the drawing. You can change the point style as many times as you wish, and points already inserted will shift to the new shape.

11.10.1 Point Style Command

- This command will set the point style. To issue this command, go to the **Home** tab, locate the **Utilities** panel, and then select the **Point Style** button:

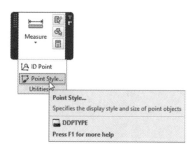

- You will see the following dialog box:

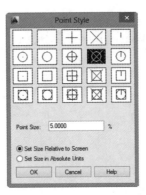

- Choose one of 20 available shapes, then set the point size, either relative to the screen or in absolute units.

11.10.2 Point Command

- This command allows you to insert as many points in a drawing that you wish.
- To issue this command, go to the **Home** tab, locate the **Draw** panel, and then select the **Multiple Points** button:

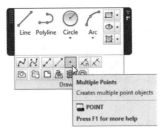

- You will see the following prompts:

```
Specify a point:
```

- AutoCAD asks you to start specifying points. Press the [Esc] key when you're done. In order to choose inserted points precisely, use the Node OSNAP.

11.11 USING THE DIVIDE AND MEASURE COMMANDS

- The Divide command will cut an object into equally spaced intervals input by you, using points, while the Measure command will cut an object into chunks with the distance you specify, using points.

11.11.1 Divide Command

- This command allows you to divide an object with equally spaced intervals specified by you. To issue this command, go to the **Home** tab, locate the **Draw** panel, and then select the **Divide** button:

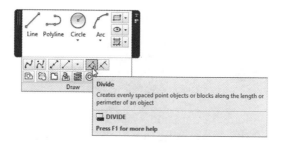

- You will see the following prompts:

```
Select object to divide:
Enter the number of segments or [Block]:
```

- AutoCAD asks you to select the desired object, then input the desired number of segments.

11.11.2 Measure Command

- This command allows you to cut an object into segments with the distance you specified, using points. To issue this command, go to the **Home** tab, locate the **Draw** panel, and then select the **Measure** button:

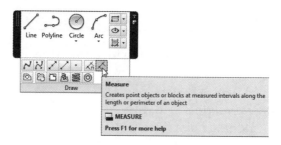

- You will see the following prompts:

```
Select object to measure:
Specify length of segment or [Block]:
```

- AutoCAD asks you to select the desired object, then to specify the desired length. When you select the object, AutoCAD will start measuring from the end nearest to the selection, so you should be careful.

11.11.3 Using the Divide and Measure Commands with the Block Option

- In both commands, you can use a block instead of a point, and the following prompts will appear:

```
Enter the number of segments or [Block]:
Enter name of block to insert:
Align block with object? [Yes/No] <Y>:
Enter the number of segments:
```

- You should first respond with the Block option to the first prompt. Then input the name of the block, with either aligned or not. Finally, enter the number of segments, or the distance depending on the command used. To understand the Aligning concept, see the following illustration:

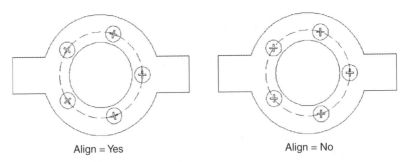

Align = Yes   Align = No

PRACTICE 11-6

Using Point Style, Point, Divide, and Measure

1. Start AutoCAD 2014.
2. Open **Practice 11-6.dwg**.
3. Change the Point Style to ⊗.

4. Zoom to the upper horizontal lines. The length of the inner horizontal line is 5700. Using the Measure command, add points at 1425 starting from the left end.
5. Erase the point at the far right.
6. Make sure that Node is on.
7. Using the Insert command, insert block = Window 1, using the three points.
8. Erase the three points.
9. Make layer Furniture current.
10. At the middle of the room, draw a circle with R = 1200. Offset the circle by 250 to the outside.
11. Using the Divide command, add the Block = Chair using the outside circle, using eight chairs.
12. Erase the outside circle. You should have something like the following:

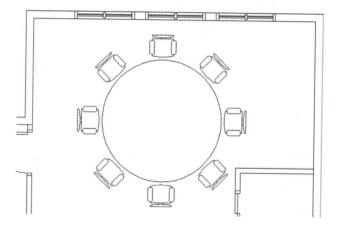

13. Save and close the file.

11.12 USING THE SPLINE COMMAND

- This command allows you to draw smooth curves based on more than two points. This command will draw a spline curve based on exact mathematical equations. There is one command, but two keys used to invoke two different methods: Fit Points or Control Vertices. To issue these two commands, go to the **Home** tab, locate the **Draw** panel, and then select one of the following two buttons:

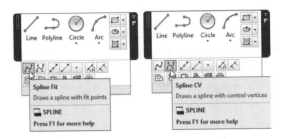

11.12.1 Fit Points Method

■ This method allows you to draw a spline with fit points coinciding with it. The following prompts will appear:

```
Specify first point or [Method/Knots/Object]:
Enter next point or [start Tangency/toLerance]:
Enter next point or [end Tangency/toLerance/Undo]:
```

■ AutoCAD asks you to specify the desired points to draw the spline with an option to close the shape automatically. We used to specify start tangency and end tangency in older versions of AutoCAD, but in this version, there is no need, as AutoCAD will determine these based on the points specified. But AutoCAD prompts also allow you to specify a start and end tangency. AutoCAD will draw a curve connecting the points you select. See the following illustration:

■ You have the ability to specify tolerance for points other than the start and end points. See the following illustration:

- AutoCAD can also convert any polyline that was created using the Polyline Edit command, and fit it in a spline to be a real spline. See the following illustration:

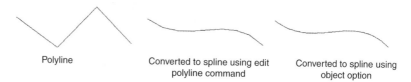

Polyline   Converted to spline using edit   Converted to spline using
                polyline command   object option

11.12.2 Control Vertices Method

- This method allows you to draw a spline using Control Vertices that define a control frame. Control frames provide a convenient way to shape the spline. The following prompts will appear:

```
Specify first point or [Method/Degree/Object]:
Enter next point:
Enter next point or [Undo]:
Enter next point or [Close/Undo]:
```

- AutoCAD asks you to specify the desired points to draw the spline with an option to close the shape automatically. You can also specify the degree of the spline that sets the polynomial degree of the resulting spline. You can input degree 1 (linear), degree 2 (quadratic), and degree 3 (cubic), and so on up to degree 10. You will get something like the following:

11.12.3 Editing a Spline

- When you click a spline created by the Fit Points method, you will see the fit points, along with the triangle, which allow you to show either the Fit Points or Control Vertices:

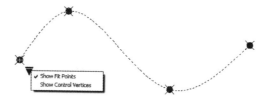

■ If you stay at one of the fit points, you will see the following menu:

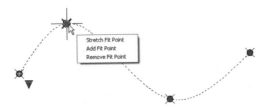

■ This menu allows you to Stretch the current fit point, Add a new fit point, or Remove the current fit point. You will see an extra option if you stay at the start or end point of the spline, which is Tangent Direction. This option allows you to change the tangent direction of the spline.

■ If you clicked a spline drawn using Control Vertices, you will see the Control Vertices, along with the triangle, which allows you to display either the Fit Points or Control Vertices:

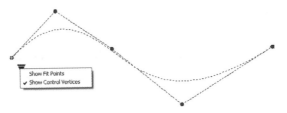

■ If you stay at one of the Control Vertices, you will the following menu:

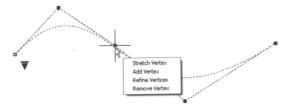

■ This menu allows you to Stretch the current vertex, Add a vertex, or Remove a vertex. The Refine Vertices option will replace the current vertex with two vertices:

PRACTICE 11-7

Using the Spline Command

1. Start AutoCAD 2014.
2. Open **Practice 11-7.dwg**.
3. Make sure Layer Contour is current.
4. Make sure that Node is on.
5. Using the points at the left draw an open spline using the Fit Points option.
6. Using the points at the middle draw a closed spline using the Control Vertices option.
7. Thaw layer Hidden Points.
8. Using grips and the Add Fit Points option add the two new points at the middle.
9. Stretch the first Fit Point (the one at the left) to the new point to its left.
10. Now the spline doesn't pass through the old first point, so using grips add it up.
11. Remove the point indicated below:

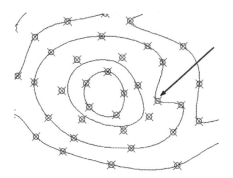

12. Freeze both points and the Hidden Points layer.
13. Save and close the file.

11.13 USING THE ELLIPSE COMMAND

- This command allows you to draw an elliptical shape or elliptical arc. To issue this command, go to the **Home** tab, locate the **Draw** panel, and then select the **Ellipse** button. You will select one of the three methods:

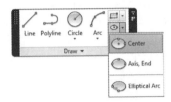

- The three options to choose from include:
  - Center
  - Axis, End
  - Elliptical Arc
- The first two options will draw an ellipse, and the third option will draw an elliptical arc. Each option is discussed as follows.

11.13.1 Drawing an Ellipse Using the Center Option

- Using this method you should specify three points:
  - Center point of the ellipse
  - Endpoint of one of the two axes
  - Endpoint of the other axis
- The following illustration shows the concept:

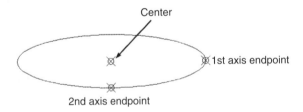

- You will see the following prompts:

```
Specify axis endpoint of ellipse or [Arc/Center]: C
Specify center of ellipse:
Specify endpoint of axis:
Specify distance to other axis or [Rotation]:
```

11.13.2 Drawing an Ellipse Using Axis Points

- You should specify three points:
  - Point on one end of one of the axes
  - Point on the other end of the same axis
  - Point on the other axis
- The following illustration shows the concept:

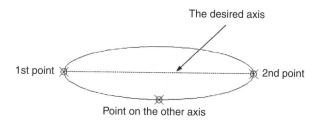

- You will see the following prompts:

```
Specify axis endpoint of ellipse or [Arc/Center]:
Specify other endpoint of axis:
Specify distance to other axis or [Rotation]:
```

- Using either method, the last step will include an option called Rotation. So, what is rotation? After you define two points, you will draw a circle. Imagine this circle is in a plane, and if the plane is rotating, you will get an ellipse. See the following illustration:

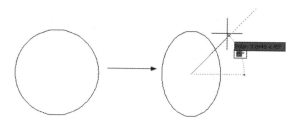

11.13.3 Drawing an Elliptical Arc

- The first three steps to draw an elliptical arc are identical to drawing the ellipse itself discussed above. Afterward, AutoCAD asks you to specify the starting angle and ending angle, counter-clockwise. Another way is after you specify the first angle, you can input the included angle and not the ending angle.

- You will see the following prompts:

```
Specify start angle or [Parameter]:
Specify end angle or [Parameter/Included angle]:
```

PRACTICE 11-8

Using the Ellipse Command

1. Start AutoCAD 2014.
2. Open **Practice 11-8.dwg**.
3. Make layer Table current.
4. Make sure Node is on.
5. Using the Axis, End method draw an elliptical table using the points displayed.
6. Freeze layer Points.
7. Make layer Window current.
8. Using Elliptical Arc, complete the window. For the outer elliptical arc, use the endpoints of the two vertical lines, and the other axis distance 3.5, then use the Offset command using distance = 0.5.
9. You should have something like the following:

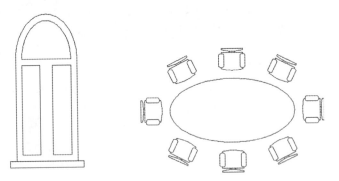

10. Save and close the file.

11.14 USING THE BOUNDARY COMMAND

- If you have several intersecting 2D objects (lines, arcs, circles, polylines, ellipses, etc.) and you want to calculate the net area of these objects, no command helps you better than this command. You can choose between polylines and regions as the resulting object. To issue this command, go to the **Home** tab, locate the **Draw** panel, and then select the **Boundary** button:

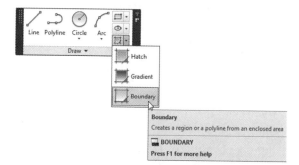

- You will see the following dialog box:

- This command depends on a simple click inside the area where you want to create a polyline. So we will start with the **choose Points** button. When you are done, click OK to end the command, and create the polyline (or region) desired. You can move it (or them) outside to calculate areas or any other desired commands. You can also make some changes to the command to get different results. See the following options:
  - Island detection: This option allows you to control whether AutoCAD should identify an object within the area.

- Polyline or Region: You can choose the desired object type.
- Boundary set: Choose whether all objects will be involved in the creation (Current viewport option) or only selected objects (click the New button).

PRACTICE 11-9

Using the Boundary Command

1. Start AutoCAD 2014.
2. Open **Practice 11-9.dwg**.
3. Zoom to Shape 01.
4. Using the Boundary command, and without changing anything, click inside the area. To see the resulting shape freeze layer Shap01.
5. Zoom to Shape 02.
6. Using the Boundary command, click off the Island detection checkbox, then click inside the area. To see the resulting shape, freeze layer Shap02.
7. Zoom to Shape 03.
8. Using the Boundary command, click the New button, and select all of Shape 03 except the four circles at the edges, then click inside the area. To see the resulting shape, freeze layer Shap03.

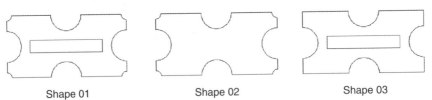

Shape 01　　　　　　　Shape 02　　　　　　　Shape 03

9. Save and close the file.

11.15 USING THE REGION COMMAND

- Assume we brought in some wires and asked you to create a rectangle and circle from the wires and to place the circle in the center of the rectangle. Though the circle is in the center of the rectangle, there is no relationship between them, because if you move one of the two shapes the other will stand still.

- On the other hand, if we brought in a piece of paper and a pair of scissors and asked you to cut a rectangle, with a hole in the shape of a circle.
- This is exactly the difference between Polyline and Region in AutoCAD.
- To issue this command, go to the **Home** tab, locate the **Draw** panel, and then select the **Region** button:

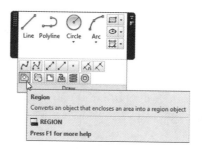

- You can create a region using the previous command Boundary.
- To convert wireframe 2D objects such as lines, arcs, circles, polylines, etc., you have to make sure they are formulating closed shapes only. You will see the following prompt:

```
Select objects:
```

- Select the desired objects, press [Enter] when done, and the objects will be converted.

11.15.1 Performing Boolean Operation on Regions

- Since regions are really 2D objects, AutoCAD can perform Boolean operations on them, including Union, Subtract, and Intersect. To issue these three commands, you have to switch to 3D Basics workspace. Go to the **Home** tab, locate the **Edit** panel, and then click one of the following buttons:

■   In the Union and Intersect commands, you can select objects in any order, but in the Subtract command, first select the region(s) you want to **subtract from**, press [Enter], and then select the region(s) to be **subtracted**.

PRACTICE 11-10

Using the Region Command

1.   Start AutoCAD 2014.
2.   Open **Practice 11-10.dwg**.
3.   Using the Region command convert all objects to regions.
4.   Click any object to show the Quick Properties and make sure that it is now a region.
5.   Using a Boolean operation, create the following shape:

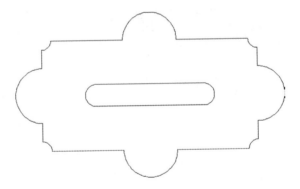

6.   Using the Properties palette, what is the total area? _____
     (19.2552)
7.   Save and close the file.

NOTES:

CHAPTER REVIEW

1. In reality donuts, polygons, revision clouds, and rectangles are all:
   a. Splines
   b. Regions
   c. Polylines
   d. None of the above.
2. In Polyline Edit, you have to select the _____ option first to join polylines to polylines.
3. Constrcution Line and Ray will produce objects as helping tools to draw accurate drawings.
   a. True
   b. False
4. You can use Boolean operations with polyline objects.
   a. True
   b. False
5. Which one of the following statements is *not* true?
   a. The Divide command will cut any object into equally spaced intervals using blocks.
   b. The Measure command will cut any object into equally spaced intervals using points.
   c. The Measure command will cut any object into chunks with the distance you specify using points.
   d. The Divide command will cut any object into equally spaced intervals using points.
6. There are two methods used to draw a spline in AutoCAD: Fit Points or Control Vertices.
   a. True
   b. False
7. The Boundary command can create either _____ or _____.

CHAPTER REVIEW ANSWERS

1. c
3. a
5. 5
7. Polyline, Region

12

ADVANCED PRACTICES – PART 1

In This Chapter

◇ The advanced features of the Offset, Trim, and Extend commands
◇ How to use Cut/Copy/Paste in multiple file
◇ How to import content from other software
◇ Hyperlink purging
◇ Views and viewport commands

12.1 OFFSET COMMAND – ADVANCED OPTIONS

- People who use AutoCAD on a daily basis often use only the default options and ignore some of the more powerful options that may help them reduce their production time significantly. In this chapter, we will look at some of the advanced options of AutoCAD.

- When you start the Offset command you will see the following prompts:

```
Current settings: Erase source=No Layer=Source
OFFSETGAPTYPE=0
Specify offset distance or [Through/Erase/Layer] <Through>:
```

- The first line is a message telling you the current values for the different AutoCAD settings: Erase source = No, Layer = Source and OFFSETGAPTYPE = 0. These settings will be discussed in the following sections.

12.1.1 Erase Source Option

■ By default AutoCAD will keep both the source and the offset object. Using this option AutoCAD allows you to keep the offset object but to erase the source object. You will see the following prompt:

```
Erase source object after offsetting? [Yes/No] <No>:
```

■ Input Yes and AutoCAD will delete the source object.

12.1.2 Layer Option

■ When you use any command in AutoCAD that produces a copy of the original object, the copy will always reside in the same layer as the source object. Using this option, you can ask AutoCAD to send the generated object to the current layer instead. AutoCAD will show the following prompt:

```
Enter layer option for offset objects [Current / Source]
<Source>:
```

■ Input Current to tell AutoCAD you want the offset object in the current layer.

12.1.3 System Variable: offsetgaptype

■ This is not an option inside the Offset command, but rather a system variable that should be invoked before the command using the Command window. This system variable decides the outcome of the shape: normal (value = 0), filleted (value = 1), or chamfered (value = 2).

12.2 TRIM AND EXTEND EDGE OPTION

■ You can't extend an object unless there is an intersecting point between the boundary edge and the objects to be extended. Also, you can't trim an object unless there is an intersecting point between the cutting edge and the objects to be trimmed. See the following illustration:

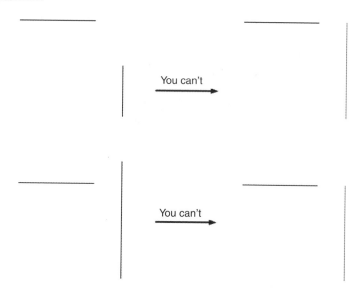

- The **Edge** option allows you to trim objects based on extended cutting edges and extend objects based on extended boundary edges. Selecting the Edge option will invoke the following prompt:

  ```
  Enter an implied edge extension mode [Extend/No extend]
  <No extend>:
  ```

- The default option is **No extend**, but you can select the **Extend** option to trim and extend based on the extended cutting and boundary edges. Note that all the above settings in Offset, Trim, and Extend affect all files from the time the change is implemented.

PRACTICE 12-1

 Using the Advanced Options in Offset, Trim, and Extend

1. Start AutoCAD 2014.
2. Open **Practice 12-1.dwg**.

3. Using the Offset command, make sure the new object will reside in the current layer and the original object will be deleted. The offset will result in a chamfered shape, using distance = 1.
4. Explode the polyline.
5. Using the Edge option in both Trim and Extend, trim the upper horizontal line using the two vertical lines and extend the lower horizontal lines based on the same vertical lines.
6. Draw small vertical lines to complete the base.
7. You should get the following shape:

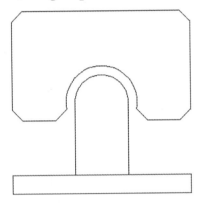

8. Save and close.

12.3 USING MATCH PROPERTIES

- This command allows you to match the correct properties of an object to the incorrect properties of other objects. Objects here include everything in AutoCAD: lines, arcs, circles, polylines, splines, ellipses, text, hatches, dimensions, viewports, and tables. To issue this command, go to the **Home** tab, locate the **Clipboard** panel, and then select the **Match Properties** button:

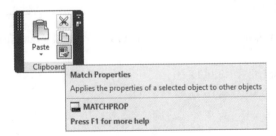

- You will see the following prompt:

```
Select source object:
```

- AutoCAD asks you to select the object that holds the correct properties. When you are done, you will see the following prompts:

```
Select destination object(s) or [Settings]:
```

- With this prompt you will see the cursor change to:

- Select the objects that have the incorrect properties, and they will be matched. You can use the **Settings** option to change the basic and advanced properties. You will see the following dialog box:

12.4 COPYING/PASTING OBJECTS AND MATCHING PROPERTIES ACROSS FILES

- In AutoCAD, you can open more than one DWG file at the same time using a simple technique used in many applications: holding the [Ctrl] key while selecting the names of the desired files in the **Open** file dialog box. But the question is why would anyone want to open more than one file at the same time? The answer would be one or both of the following:
  - To copy objects from one file to another
  - To match properties across files
- To tile the opened files, go to the **View** tab, locate the **Windows** panel, and then use one of the following two buttons:

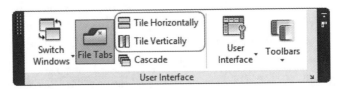

- We will use the normal Copy/Paste sequence in order to copy objects from one file to another (this Copy command is different than the Copy command in the Modify panel, as this command will copy objects from one file to another). Take the following steps:
  - Without issuing any command, select the desired object(s).
  - Right-click and select the Clipboard option and then select one of the two copying commands available.
  - Go to the other file. Right-click and select the Clipboard option and then select one of the three available pasting commands.
- These are the **Clipboard** options:

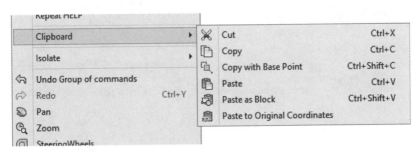

12.4.1 Copying Objects

- AutoCAD allows you to copy objects from one file to another using two techniques:
    - The **Copy** option allows you to copy objects without specifying a base point.
    - The **Copy with Base Point** option allows you to copy objects by specifying a base point. You will see the following prompt:

    ```
    Specify base point:
    ```

12.4.2 Pasting Objects

- There are three methods to paste objects across files:
    - The **Paste** option allows you to paste the contents of the clipboard.
    - The **Paste as Block option** allows you to paste objects as a block with an arbitrary name. You can use the Rename command to give the block a specific name.
    - The **Paste to Original Coordinates** option allows you to paste objects to the same coordinates used in the original file.

12.4.3 Using the Drag-and-Drop Method

- You can also use the Drag-and-Drop method to copy and paste. Using this method, you select the desired objects in the source file, then click and hold the left button and go to the destination file and drop the objects there. You can also do this with the right mouse button, but when you drop the objects in the destination file, a menu will appear as follows:

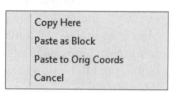

- This menu includes self-explanatory options.
- While you are copying objects across files AutoCAD will create the necessary layers, text styles, dimension styles, etc.

12.4.4 Matching Properties Across Files

- You will use the same command to match properties across files selecting the source object in the current file (the file you will issue the command from) and then matching the objects holding the incorrect properties in the destination file. AutoCAD will also create the necessary layers, text styles, dimension styles, etc., in the destination file using this method.

PRACTICE 12-2

Matching Properties, Copying/Pasting Across Files

1. Start AutoCAD 2014.
2. Open **Practice 12-2.dwg** and **Ground Floor.dwg**.
3. Tile them vertically.
4. Look at the layers in Practice 12-2.dwg, and a note the existing layers.
5. Look at the dimension styles in Practice 12-2.dwg, and note the existing styles.
6. Using Copy/Paste copy the bathroom 9' × 9' from Ground Floor.dwg using the base point to a suitable place in Practice 12-2.dwg.
7. Using Match Properties select any dimension in the Ground Floor.dwg and match it with all the dimensions in Practice 12-2.dwg.
8. Close Ground Floor.dwg and maximize Practice 12-2.dwg.
9. In Practice 12-2.dwg, match the properties of the copied text with all the other texts.
10. Start the Match Properties command and select the hatch of the kitchen as the source object, then select the hatch of the two toilets.
11. Look at the layers again. See how many layers were added?
12. Look at dimension styles again. See how many new styles added?
13. If you have time, put all the objects in the right layer.
14. Save and close the file.

12.5 SHARING EXCEL AND WORD CONTENT IN AUTOCAD

- As an application running under Windows, AutoCAD can take and give content to and from any other Windows application, especially MS Office software such Word and Excel, the most commonly used software today.

AutoCAD will use OLE (**O**bject **L**inking & **E**mbedding) to copy the content from and to AutoCAD. The Paste command used in AutoCAD will dictate the type of object brought into AutoCAD. Sharing content between Windows applications is a common practice and easy to do.

12.5.1 Sharing Data From Word

- Copy the content from Word using Copy or Cut, then go to the **Home** tab and locate the **Clipboard** panel, as shown below:

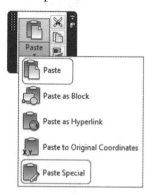

- We will use either the Paste or Paste Special options. See the following:
  - Word Content ➜ Using Paste ➜ OLE Object Embedded
  - Word Content ➜ Using Paste Special / Paste / Text ➜ MTEXT
  - Word Content ➜ Using Paste Special / Paste / UniCode Text ➜ MTEXT
  - Word Content ➜ Using Paste Special / Paste / AutoCAD Entities ➜ Text
  - Word Content ➜ Using Paste Special / Paste Link ➜ OLE Object Linked
- Once you issue the **Paste Special** command, you will see the following dialog box:

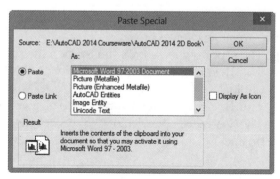

■ As you can see the first option (using Paste only) and the last option (using Paste link) will bring in an OLE object. Using the first method, the content will not be updated, but the second method will update the content as needed. Using the first method, you will see the following dialog box:

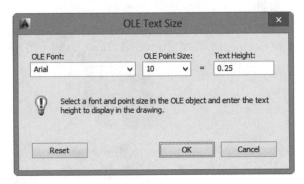

■ The OLE Text Size dialog box will show the font name and font size used in Word and then asks you to specify the **Text Height** in AutoCAD units. You can click the Reset button at any time to go back to the original values. Once you are done, click **OK**.

■ After pasting the OLE object, AutoCAD allows you to edit it. To edit an OLE object, select it, right-click, and you will see a menu. Select the **OLE** option, and you will see the following:

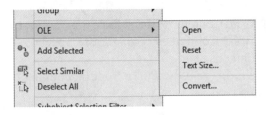

■ The available choices are:
  • The **Open** option allows you to open the source application.
  • The **Reset** option allows you to retain the font and font size of the original.
  • The **Text Size** option allows you to change the text size.
  • The **Convert** option allows you to change the nature of the OLE object.

12.5.2 Sharing Data From Excel

- You can copy content from Excel using Copy or Cut, then go to the **Home** tab, locate the **Clipboard** panel, and choose either the Paste or Paste Special option, like we did in Word. In Excel, you have the following choices:
  - Excel Content ➔ Using Paste ➔ OLE Object Embedded
  - Excel Content ➔ Using Paste Special / Paste / Text ➔ MTEXT
  - Excel Content ➔ Using Paste Special / Paste / UniCode Text ➔ MTEXT
  - Excel Content ➔ Using Paste Special / Paste / AutoCAD Entities ➔ Table Embedded
  - Excel Content ➔ Using Paste Special / Paste Link ➔ OLE Object Linked
  - Excel Content ➔ Using Paste Special/Paste Link/AutoCAD objects ➔ Table Linked
- You will see the same dialog boxes discussed above.
- The last choice allows you to paste a linked table in AutoCAD and will be discussed in the following.

12.5.3 Pasting a Linked Table From Excel

- This option allows you to paste Excel content into your current drawing, linking it to the original Excel sheet. With this option, when you update the file in Excel, the table in AutoCAD will be updated, and vice-versa – when you update the table in AutoCAD it will affect the file in Excel. When you hover over the table inside AutoCAD you will see the following image:

- This means the table inside AutoCAD is *locked* and *linked*.
- When you change anything in the Excel sheet and save, these changes will be reflected in the AutoCAD table. At the lower-right corner of the AutoCAD window, you will see a chain called **Data Link** just like the following:

- If you right-click this icon, you will see a menu just like the following:

- Select the **Update All Data Links** option to get the latest copy of your Excel sheet. If the AutoCAD file is not opened, the next time you open it, you will get the newest copy of the Excel sheet.
- Updating the Excel content from the AutoCAD table is a little bit more complicated.
  - Go to the **Insert** tab, locate the **Linking & Extraction** panel, and then select the **Data Link** button:

  - The following dialog box will come up:

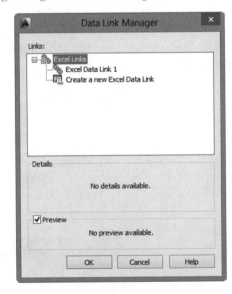

- Under **Links** you will see there is a link called **Excel Data Link1**, which was created automatically when you pasted the link for the Excel table. This name is temporary. To rename the link, click it, right-click, and then select the **Rename** option:

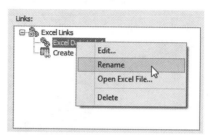

- There will be a **Preview** checkbox turned on to help you preview the link in case there is more than one link in the same file.
- Double-clicking the link or selecting the **Edit** option from the right-click menu will lead to the following dialog box:

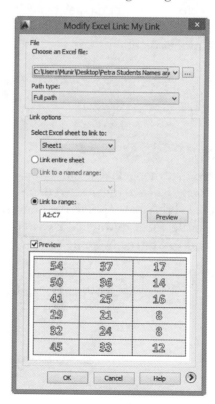

- Clicking the arrow at the lower-right corner of the dialog box will expand it to show more options:

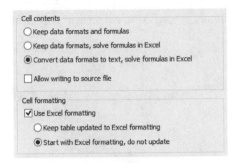

- Make sure that the **Allow writing to source file** checkbox is turned on because makes sure your Excel sheet stays updated.
- Click **OK**, several times to close all dialog boxes.
- We are ready to make the edits in the table pasted in AutoCAD.
- Using crossing, select the desired cells in the table.
- Right-click and a menu will appear. Select the **Locking** option, then the **Unlocked** option:

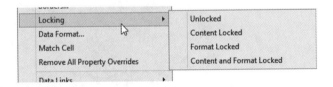

- Notice that the locking symbol disappeared, but the chain symbol is still there. Change the desired value in the unlocked cells.
- To upload these changes to the original Excel sheet go to the **Insert** tab, locate the **Linking & Extraction** panel, and then select the **Upload to Source** button:

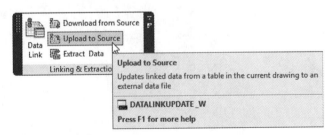

- AutoCAD will ask you to select objects. Click the desired table from one of its outside borders.
- The following prompt will appear:

```
1 object(s) found.
1 data link(s) written out successfully.
```

- The following bubble will appear at the lower-right corner of the AutoCAD window:

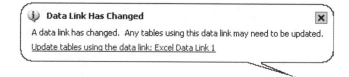

PRACTICE 12-3

Sharing Excel and Word Content in AutoCAD

1. Start AutoCAD 2014.
2. Open **Practice 12-3.dwg**.
3. Using Word, open the file House General Notes.doc and copy all its contents.
4. Go to Full Plan layout.
5. Make layer Text current.
6. Using the Paste option, paste the text in the lower-left corner of the layout below the viewport.
7. Make the text size = 5.
8. Open the file Door Cost Schedule.xls and copy its contents.
9. Using Paste Special/Paste Link, select the AutoCAD entities.
10. Close the Excel sheet.
11. Change the quantity of Type 02 to 21 and updated the Excel sheet.
12. Update the AutoCAD file with the new value to recalculate the new values.
13. Make sure that the Excel file has changed.
14. Save and close the files.

12.6 HYPERLINKING AUTOCAD OBJECTS

- This command allows you to hyperlink any AutoCAD object(s) to a website, an AutoCAD drawing, an Excel sheet, a Power Point presentation, etc. To issue this command, go to the **Insert** tab, locate the **Data** panel, and then select the **Hyperlink** button:

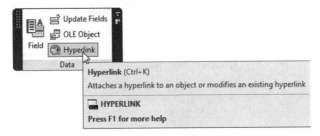

- You will see the following prompt:

```
Select objects:
```

- Select the desired objects. When done press [Enter], and you will see the following dialog box:

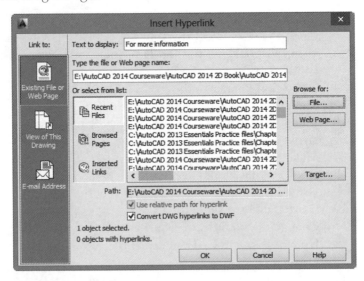

- Fill in the **Text to display**, which is like a helpful note that appears when you get close to the object holding the hyperlink. Then **Type the file or Web page name**, which is a Web site address or a file path. Finally, click **OK** to end the command.

- Using the **Home** tab, locate the **Clipboard** panel, and you will find an option called **Paste as Hyperlink**, which will do the job in reverse and will paste the contents as a hyperlink to an object.

PRACTICE 12-4

Hyperlinking AutoCAD Objects

1. Start AutoCAD 2014.
2. Open **Practice 12-4.dwg**.
3. Hyperlink the 3D shape to a file called Part Detail Dimension.dwg.
4. Test the hyperlink by holding the [Ctrl] key and clicking the 3D shape.
5. Save and close the files.

12.7 PURGING ITEMS

- We normally create lots of content inside AutoCAD drawings, such as layers, blocks, dimension styles, text styles, multileader styles, etc. We use some of it, and other content is left unused. The Purge command will help you get rid of this unused content. To reach this command, go to the **Application** menu, select **Drawing Utilities**, and then select the **Purge** command:

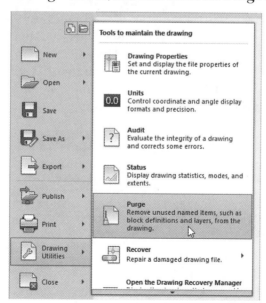

- You will see the following dialog box:

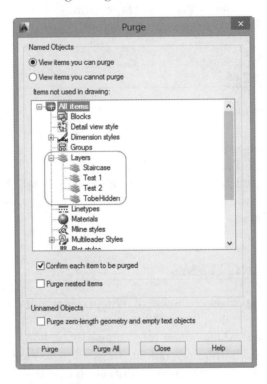

- AutoCAD will list all the unused items. In the above example, you can see that there are two unused layers and one unused multileader style. You will also see three checkboxes:
  - Confirm each item to be purged: If is this turned on, you have to decide to purge or not each time you want to purge an item. You will see something like the following:

- Purge nested items: This option will help you purge nested blocks. A nested block is a block that contains blocks. If this option is turned off, you will purge only big blocks, and the nested blocks will remain. But if you turn it on, AutoCAD will purge the big blocks and the nested blocks, as long as they are not being used in the current file.
- Purge zero-length geometry and empty text objects: This option will remove all lines, arcs, and polylines that have a length of zero, and all MTEXT, text that contains only spaces.

■ Finally, you can purge one item at a time, or purge all items in one step. Select either button then click the **Close** button to end the command.

■ You can't purge the current layer, text style, dimension style, etc.

■ The Purge command can reduce the file size significantly.

PRACTICE 12-5

 Purging Items

1. Start AutoCAD 2014.
2. Open **Practice 12-5.dwg**.
3. Start the Purge command and type the name of the layers to be purged:
   _____, _____, _____, _____
4. Click the plus sign beside the Layers category, then select Staircase and click the Purge button. The Purge this item dialog box will appear.
5. Repeat the same for the other layers.
6. Click Purge All to purge all the other items, and select the Purge All Items option.
7. See if there are any plus signs beside any other items; if not, click the Close button.
8. Save and close the files.

12.8 USING VIEWS AND VIEWPORTS

12.8.1 Creating Views

- A view in AutoCAD is any rectangular shaped portion of a drawing that can be saved under a name. There are two ways to define it: either by zooming to the part of the drawing you want to save, then issuing the command, or by using the View command. To issue this command, go to the **View** tab, locate the **Views** panel, and then select the **View Manager** button:

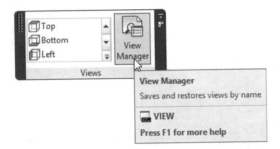

- You will see the following dialog box:

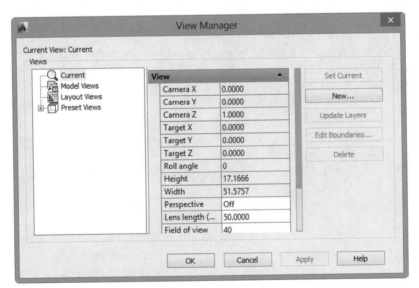

- To create a new view, click the **New** button, and the following dialog box will appear:

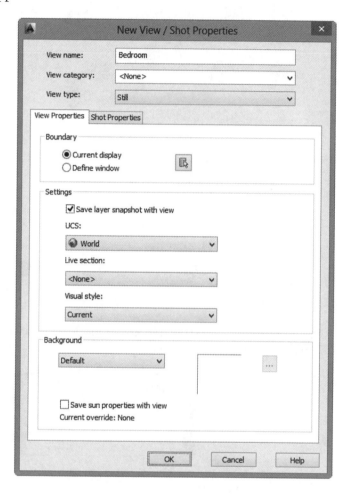

- As a first step, input the **View name**. Under **Boundary** select whether your view is the **Current display**, or if you want to **Define window** (click the small button at the right to zoom to the desired area). Using **Settings**, select whether you want to **Save layer snapshot with view**. The layer snapshot is the current status of layers (on/ off, Thaw/Frozen, Unlock/Lock, etc.). When done click **OK** to save the view.

- There are multiple ways to retrieve the saved views:
  - You can use the In-canvas View control at the upper-left corner of the screen, using **Custom Model Views** as shown below:

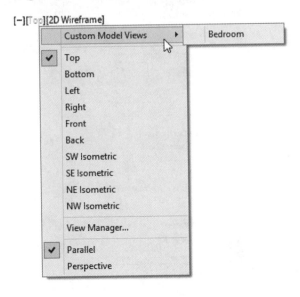

  - You can use the **View** tab, locate the **Views** panel, and then click the name of the desired view, as shown below:

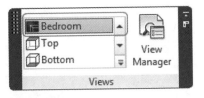

12.8.2 Using Views in Viewports

- You can show the saved views in viewports using the **Viewport** dialog box. Start the **Viewport** dialog box (if you are in one of the layouts, you have to use Vports or the Viewports command). Select one of the arrangements (the example below is Three: Right) using Preview. If you click on one of the viewports it will be current, and you can select one of the saved views to show inside it:

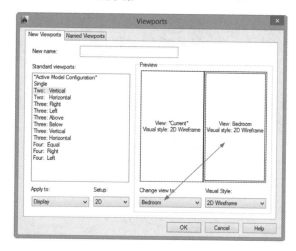

12.8.3 Creating a Named Viewport Arrangement

- If you don't like the viewport arrangement AutoCAD provides, and you are thinking about creating your own arrangement, this section is for you. Take the following steps:
  - Make sure you are in Model space.
  - Go to the **View** tab, locate the **Model Viewports** panel, and then select the **Viewport Configuration** button to see something like the following:

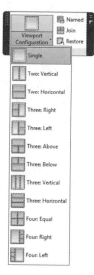

- Or you can reach the same command using the In-canvas viewport control:

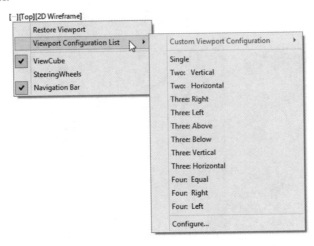

- Choose one of the existing arrangements to begin with.
- Select one of the Model viewports.
- Start Viewport command again, and pick an arrangement.
- Make sure that at the lower-left corner of the dialog box, under Apply to, you select Current Viewport, then click **OK**:

- This should make your small viewport even smaller.
- Do the same to different viewports.
- Go to the **View** tab, locate the **Viewports** panel, and select the **Join** button. This allows you to join adjacent viewports (the condition here is to form a rectangular shape):

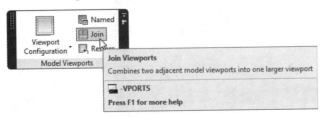

- Once you are done joining viewports, you will end up with a new arrangement, and all you have to do is save it. Start the Viewport command and type the name of the new arrangement as shown below:

- To retrieve it in the layout, go to the desired layout.
- Go to the **View** tab, locate the **Viewports** panel, and click the **Named** button:

- Choose the desired named viewport arrangement and insert it as normal.

PRACTICE 12-6

Using Views and Viewports

1. Start AutoCAD 2014.
2. Open **Practice 12-6.dwg**.
3. Create four views, making sure that you save a layer snapshot with the view. The four views are:
    a. Master Bedroom
    b. Master Bedroom Bathroom
    c. WIC
    d. Kitchen
4. Thaw layer Dimension.
5. Retrieve one of the four views. What happened to layer Dimension? Why _____

6. Insert Four: Equal viewports in the Detail 1 layout showing a view in each viewport.
7. Using Model space create Four: Equal.
8. For the upper-left one, split it into Two: Vertical (using the current viewport option).
9. For the lower-right one, split it into Two: Vertical (using the current viewport option).
10. For the upper ones join the small to big ones and for the lower ones join the small one to the bigger one, then save this new arrangement under the name Four Unequal.
11. Go to the Detail 2 layout and insert the new arrangement. Going to each viewport zoom to different parts of the floor plan.
12. Save and close the files.

NOTES:

CHAPTER REVIEW

1. Which one of the following statements is *not* correct?
   a. Using the Paste option, the content will always be an OLE object in AutoCAD.
   b. You can insert a table in AutoCAD from an Excel sheet and the editing will be updated both ways.
   c. When you paste content from Word, it will always be MTEXT.
   d. To link the content you have to use Paste Special.
2. Use _____ in the Offset command to create filleted or chamfered edges while offsetting.
3. All of the following is true about the View and Viewport commands *except*?
   a. You can save a layer snapshot with a view.
   b. You can create a new viewport arrangement in Model and Paper spaces.
   c. You can insert a new viewport arrangement in Paper space.
   d. You can show a saved view in each viewport.
4. You can match properties across files.
   a. True
   b. False
5. While copying objects between files you can use only drag-and-drop using the left mouse button.
   a. True
   b. False
6. The_____ option allows you to trim objects based on extended cutting edges.
7. Using the right-click menu and the Clipboard option, you can use the Paste Special option.
   a. True
   b. False

CHAPTER REVIEW ANSWERS

1. c
3. b
5. b
7. b

Chapter 13 — ADVANCED PRACTICES – PART 2

In This Chapter

- ◇ Using the Content Explorer
- ◇ Working with fields
- ◇ Formulas and tables and advanced table functions
- ◇ Quick Select, Select Similar, and Add Selected commands
- ◇ Partial Open and Partial Load
- ◇ Object visibility

13.1 USING AUTODESK CONTENT EXPLORER

- ■ Unlike Design Center, using Autodesk Content Explorer you can search drivers and folders containing thousands of AutoCAD files for anything you want. Autodesk Content Explorer's Google-like index allows it to search and find the desired content quickly. To issue this command, go to the **Plug-Ins** tab, locate the **Content** panel, and then click the **Explore** button:

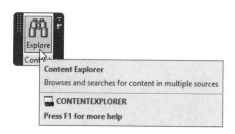

- You will see the following palette:

- By default, AutoCAD will look in two folders and one online site. The two folders are the Sample folder that comes with AutoCAD and the Downloaded Content folder. The online site is Autodesk Seek. In order to add more folders, click the **Configure Settings** button:

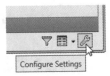

- You will see the following dialog box:

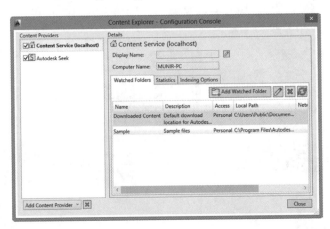

- Click the **Add Watched Folder** button to add more folders to the search process.
- To start a search, simply type your word(s) in the **Search** field, like the following:

- Then click the magnifying glass, and the results will be listed instantly:

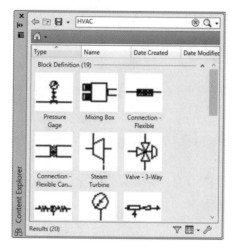

- In this example, AutoCAD lists one file and three pieces of multiline text in it that contain the words precast concrete. When you double-click any of these three multiline objects, AutoCAD will open the file and zoom to the text.
- The two buttons at the bottom will help you **Toggle icon size**, you will see the following menu:

- The last button is **Configure filtering**, which will help you filter the things that AutoCAD will search for when you input the search phrase. You will see the following menu:

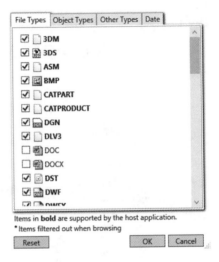

- Specify the **File Types**. Click the **Object Types** tab, and you will see:

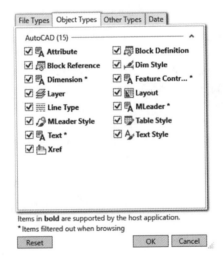

- Specify what objects AutoCAD should look for. Click the **Other Types** tab, and you will see:

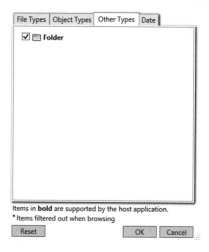

- Click the **Date** tab to see the following:

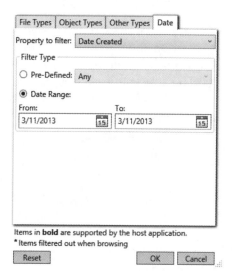

- Specify whether AutoCAD should search for all files created at anytime, or specify a date range for searching.

PRACTICE 13-1

Using Autodesk Content Explorer

1. Start AutoCAD 2014.
2. Open **Practice 13-1.dwg**.
3. Start Autodesk Content Explorer.
4. Make sure that the Sample folder that comes with AutoCAD is included; if not, add it as a Watched folder.
5. Search for word **duct**.
6. The result should be two files, which contain four leaders and two pieces of multiline text.
7. Double-click the Mtext that starts with the word "NOTES DUCT-WORK: ...."
8. Copy it and paste it in your file.
9. Save and close both files.

13.2 USING QUICK SELECT

- The Quick Select command allows you to select object in the current drawing based on their properties, which is handy in dense and complicated drawings containing hundreds of thousands of different types of objects. There are multiple ways to issue the Quick Select command:
  - Go to the **Home** tab, locate the **Utilities** panel, and then select the **Quick Select** button:

  - Without issuing any command, right-click, and then select the **Quick Select** option from the menu:

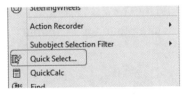

- You will see the following dialog box:

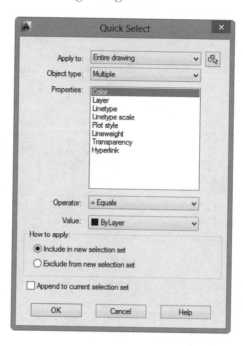

- This method is similar to using SQL (Structured Query Language) and will apply a filter to find information-based properties.
- First, select the **Apply to** part: either Entire drawing or click the button beside it to filter by an area of your choice. Select the **Object type** to search for. AutoCAD will list objects found in the current file, and you will see the following:

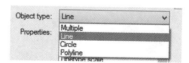

- Select the desired object and then select the **Properties** of the selected object. If you select **Multiple** for **Object type**, you will see only general properties; if you select an object type, then you will see both general and specific properties to choose from. The following is the list of properties of a circle:

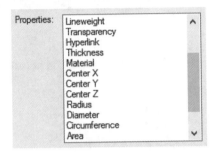

- The next step is to select the desired **Operator**. You will see something like the following:

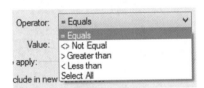

- Specify the proper **Value** for the selected operator.
- Finally, select **How to apply** the filter in the drawing. You can create a fresh new selection set from it, or if you selected objects prior to the Quick Select command, you can exclude the objects from it, or you can append it to already selected objects.
- As a final note, you can access Quick Select while you are at the Properties palette, just like the following:

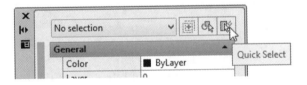

PRACTICE 13-2

 Using Quick Select

1. Start AutoCAD 2014.
2. Open **Practice 13-2.dwg**.
3. Since the drawing says that we have 12 × R 0.06 we want to make sure that this is right. So using Quick Select, select all circles with R = 0.06. How many circles are selected? _____
4. Using Quick Select and Properties, change the radius of all the other 11 circles to have R = 0.06.
5. Using Quick Select and Properties move all circles of R = 0.06 to layer Holes.
6. Using Quick Select and Properties move all dimensions to layer Dimensions. (Hint: Linear dimensions are called Rotated dimensions, and since we have two types, rotated and radial, you should use append, or you can do this step twice.)
7. Save and close the file.

13.3 USING SELECT SIMILAR AND ADD SELECTED

- The Similar command allows you to select an object and then select all similar objects that have the same properties. The Add Selected command allows you to select an object and then initiate the command that will draw the same object with the same properties.

13.3.1 Select Similar Command

- There are two ways to issue this command. Using the first method you will:
  - Select the desired object.
  - Right-click, then choose the **Select Similar** option:

- Based on the current settings, AutoCAD will select the similar objects.

- The second method involves the following steps:
  - At the Command window type SELECTSIMILAR, and the following prompt will appear:

    ```
    Select objects or [SEttings]:
    ```

  - Right-click and select the **Settings** option, and you will see the following dialog box:

- As shown above, AutoCAD will select based on layer and name. Other things to choose from are Color, Linetype, Linetype scale, Lineweight, Plot style, and Object style (anything that has a style such as dimension, text, etc.). If you select more than one object (arc and circle), then right-click and choose **Select Similar**, and AutoCAD will select objects similar to both selected objects.

13.3.2 Add Selected Command

- Let's assume you have issued a command that includes layer Centerline, Color yellow, and Linetype dashdot, and you want to create a new line that lies in the same layer and holds the same properties. This command will help you accomplish your mission easily. Select the desired object, then right-click and choose the **Add Selected** option:

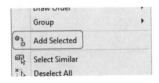

- AutoCAD will start the command, and you are ready to start drafting. The new object will reside in the same layer, and it will hold the same properties as the selected object.

13.4 WHAT IS OBJECT VISIBILITY IN AUTOCAD?

- The normal practice in AutoCAD to hide an object is to turn the layer off, or freeze it. But this means all the other objects in the same layer will be hidden as well. This command allows you to hide an object or group of objects without hiding all the other objects in the same layer. To do that select the desired object (or as many objects as you wish), right-click, and then select the **Isolate** option. You will see the following:

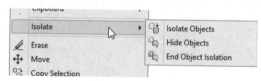

- As you can see there are three commands: Isolate Objects, Hide Objects, and End Object Isolation:
  - Isolate Objects will show the selected objects and will hide all other objects.
  - Hide Objects will hide the selected objects and show all other objects.
  - End Object Isolation will cancel the first command, i.e., it will show all objects.
- You will see a bulb at the lower-right corner of the screen; it is either on or off. If you click it, you will see the same menu, so you can do everything from there as well:

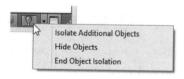

PRACTICE 13-3

Using Select Similar and Add Selected

1. Start AutoCAD 2014.
2. Open **Practice 13-3.dwg**.
3. Type SELECTSIMILAR and select the Settings option. Clear all checkboxes and then select one of the red circles. Press [Enter].
4. You will notice that AutoCAD selected all circles in the drawing.
5. Right-click and select Isolate/Isolate Objects.
6. Make the Centerlines layer current.
7. Start the Line command. Using OTRACK and the Center point of one of the two large circles take 1.2 from the center to the left to start the line, then draw a 2.4 horizontal line that passes through the center of the two large circles.
8. End Object Isolation.
9. Make layer 0 current.
10. Select one of the green centerlines at the right or at the left, right-click, select Add Selected, and draw a vertical centerline for the large circles.
11. Does it look different than the horizontal line? _____ If so, why? (The linetype scale of the two lines at the right and the left is 0.5, so the new line holds all the properties of the original line.)
12. Select one of the two linear dimensions, then right-click and choose Select Similar. Two dimension blocks will be selected, right-click, select Precision, and select 0.000.
13. Select the four centerlines, right-click, and then select Isolate/Hide Objects.
14. Save and close the file.

13.5 ADVANCED LAYER COMMANDS

- In this section, we will discuss some advanced commands related to layers. Most of these commands depend on selecting a tool to perform a certain task related to the layer of the object selected. These commands make our lives easier and help save time. These commands can be found at the **Home** tab, in the **Layers** panel:

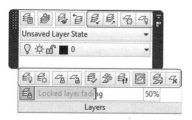

13.5.1 Using the Isolate and Unisolate Commands

- Use the following two buttons:

- The Isolate command will ask you to select object(s), and the layer(s) of these objects will be shown, while the other layers will be turned off (or locked). The Unisolate command, on the other hand, cancels the effects of the Isolate command.

13.5.2 Using the Freeze and Off Commands

- Use the following two buttons:

- The Freeze and Off commands will ask you to select object(s), and the layer(s) of these objects will be turned off, or frozen.

13.5.3 Using Turn All Layers On and Thaw All Layers Commands

- Use these two buttons:

- These two commands will turn all layers on and will thaw all layers, a very handy tool to do this process in one step.

13.5.4 Using the Lock and Unlock Commands

- Use these two buttons:

- The Lock command will ask you to select the object(s), and the layer(s) of these object(s) will be locked. You can lock one layer at a time. The Unlock command will ask you to select object(s), and the layer(s) of these objects will be unlocked. You can unlock one layer at a time.

13.5.5 Using the Change to Current Layer Command

- Use this button:

- This command will ask you to select object(s), and then it will change the layer of these object(s) to be the current layer.

13.5.6 Using the Copy Objects to New Layer Command

- Use this button:

- This command will ask you to select object(s), then it will copy them to a new location in the drawing, and then you can change the layer of the new objects to a new layer either by selecting an object in the desired layer, or by typing its name.

13.5.7 Using the Layer Walk Command

- Use this button:

- This command will show a dialog box listing all the layers in the current file. To show the contents of a layer, click its name in the list (by default all layers are selected). A checkbox at the bottom says "Restore on Exit," which means, whenever you close this dialog box, all the layers will restore to their previous status. If this layer is turned off, this command will keep its effects:

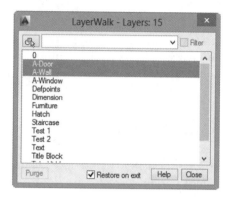

13.5.8 Using the Isolate to Current Viewport Command

- Use this button:

- This command will ask you to select object and then freeze the layer of this object in all viewports except the current viewport.

13.5.9 Using the Merge Command

- Use this button:

- This command allows you to merge a layer or more in a target layer. First, AutoCAD will ask you to select an object (AutoCAD will list the name of the layer of the selected object), and AutoCAD will keep asking you to select objects until you press [Enter] and are done. The last step will be to select an object in the target layer, and AutoCAD then will report the deleted layers.

13.5.10 Using the Delete Command

- Use this button:

- This is a very handy command, as we know that AutoCAD will not delete a layer unless it is empty. This command will delete (purge) a layer by selecting an object that resides in it, except the current layer.

13.6 LAYER TRANSPARENCY

- You can set the visibility of a layer. The default value for all layers is 0 (zero) and can be as much as 90. If you are making a test plot for a drawing with lots of solid hatching, set the visibility as the minimum so you don't use too much plotter ink. You can use the same color for lots of layers, then control the visibility of the layer to give each layer a different tone of the color. When you start the Layer Properties Manager, you will see a column called Transparency, and if you are at a layout, you will see another column called VP Transparency:

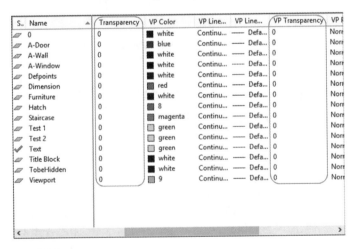

- You can control the visibility of new objects. To do that go to the **Home** tab and locate the **Properties** panel:

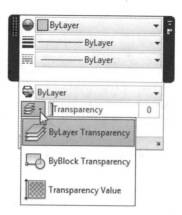

- Select either to set the **Transparency** to be ByLayer, ByBlock, or Transparency Value (which will set it to 0 (zero)) or move the slider to any desired value.
- Also, set the value for object(s) using the Properties palette, as follows:

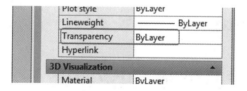

- To see the effect of Transparency use the Status bar to show or hide it:

PRACTICE 13-4

Using Advanced Layer Commands

1. Start AutoCAD 2014.
2. Open **Practice 13-4.dwg**.
3. Using the Freeze command select one of the dimension blocks. What happened to the Dimension layer? _____
4. Start the Isolate command and select a line of the walls. What happened to the other objects in the other layers? _____
5. Unisolate.
6. Start the Layer Walk command. You will notice that all layers are selected.
7. Click any of the layer names, select the A-Door layer, hold [Ctrl], and select A-Wall and A-Window. Uncheck the Remove on exit checkbox and click Close. At the warning message click Continue.
8. Start the Layer Properties Manager palette. What is the state of the layers you didn't select in the previous step? _____
9. Start Layer Walk again, and while holding the [Ctrl] key, select the Furniture and Toilet Furniture layers. Close the Layer Walk command.
10. Merge layer Toilet Furniture with layer Furniture.
11. What happened to the Toilet Furniture layer after the merging process?
12. Save and close the file.

13.7 USING FIELDS IN AUTOCAD

- AutoCAD stores lots of data in the drawing database. Some of it is constant and some of it is variable. You can utilize these types of data by inserting them in drawings to benefit from the updating feature that AutoCAD performs on the variable data. This approach is better than writing text using the Text and Mtext commands, which will need to be updated manually. There are four methods to insert a field in your drawing:

- Using the Field command. To issue this command, go to the **Insert** tab, locate the **Data** panel, and then select the **Field** button:

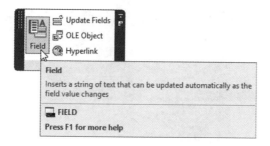

- Using the Text command, and after you specify the starting point, height, and rotation, and before you start writing, right-click, and select the **Insert Field** option:

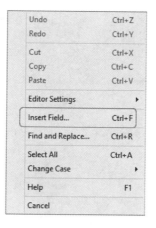

- Using the Mtext command, and after you specify the text area, you will find at the **Text Editor** context tab the **Insert** panel; select the **Field** button:

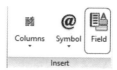

- You can also insert fields in a table and use the Attribute commands.

- Regardless of the method used to reach the **Field** command, the following dialog box will appear:

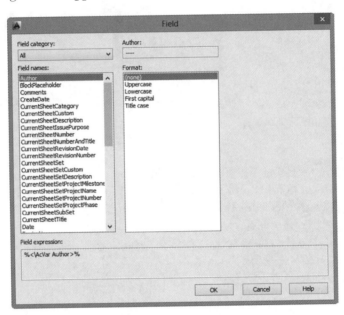

- The first thing to control is the Field category, which will help you find the desired data quickly. Clicking the pop-up list will show the following list:

- There are seven field categories, and each will show the related field names. For instance, the Objects category will show four field names:

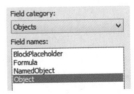

- Based on the field category and field name, you will find at the right side of the dialog box things to help you control the appearance of the field. For instance, if you select the Field category Document and the Field name is File name, you will see the following at the right:

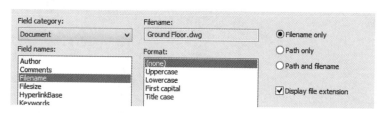

- As you can see, you will select the Format and whether you want to show Filename, Path only, or Path and filename. Finally, you can control whether to show the file extension. If you use Field category Objects and Field name Object, you will need to select an object from the drawing:

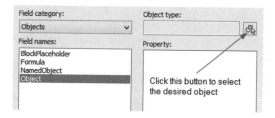

- When you click this button, the dialog box will disappear temporarily to let you select the desired object. When done, you will see something like the following:

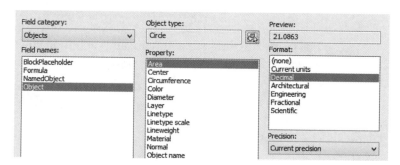

- As you can see, we selected a circle, and we chose to use the Area with decimal format and 0.00 precision.

- When you insert a field in the drawing, the default settings will display it with a background, just like the following:

- To remove the background, go to the Application menu and select the **Options** button. Select the **User Preferences** tab, then locate **Fields**, and uncheck the **Display background of fields** checkbox, just like the following:

- Click the checkbox **Display background of fields** off if you don't want to display the background. Click **Field Update Settings** to see the following dialog box:

- Control to update the field automatically each time you save, open, plot, eTransmit, or Regen. But if you need to update manually, then go to the **Insert** tab, locate the **Data** panel, and then select the **Update Fields** button:

- You need to select the desired fields to be updated.

PRACTICE 13-5

Using Fields in AutoCAD

1. Start AutoCAD 2014.
2. Open **Practice 13-5.dwg**.
3. Make sure that the current layer is Polyline.
4. Using the Boundary command, click inside the master bedroom to add a polyline.
5. Make layer Text current.
6. Go to the Annotate tab and make sure that Room Titles is the current text style.
7. Start the Single Line text command and specify a point almost at the middle of the room, with rotation angle = 0. Type "Area", = then right-click and select the Insert Field option and insert the area of the polyline showing 266 SQ. FT.
8. Go to Full Plan layout.
9. Zoom to the lower-right of the title block.
10. Using the Application Menu, go to Drawing Utilities, then select Drawing Properties. Go to the Summary tab, and at the Title input Munir Hamad Villa, at Author input your name. Click OK to end the command.
11. Make Standard the current text style.
12. Using the Field command, and under the title Project Name, insert a Title field using uppercase letters. Under Designed By insert Author, using an initial capital letter.
13. Under Date insert today's day using the MMMM d, yyyy format.
14. Under Filename insert the filename only without the extension using lower case letters.
15. Remove the background for fields.
16. Save and close the file.

13.8 USING PARTIALLY OPENED FILES

- At some point, you will probably work with a huge drawing that contains many views and layers. Big files tend to take a long time to open. To eliminate this problem, you can use the Partial Open command. Later, you can use the Partial Load command to add more content to the partially opened file.

13.8.1 How to Partially Open a File

- Use the normal Open command and select the desired file, but DON'T click Open as you always do, instead click the small arrow at its right to see the following list of options. Select **Partial Open**:

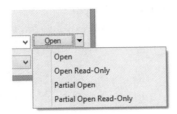

- You will see the following dialog box:

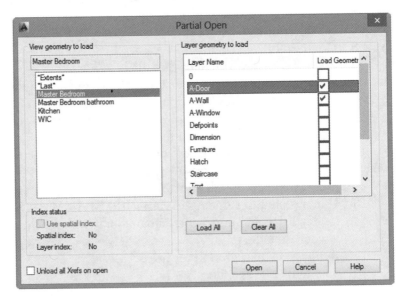

- You can do all or any of the following:
  - Select the desired layers to open
  - Select the desired views to open, or you can use Extents or Last views
  - Then click Open to open the file with the settings you selected

13.8.2 Using Partial Load

- This command is not applicable for normal files and is only used for files that are partially opened. To use this command, make sure the menu bar is shown, then select **File/Partial Load**, and you will see the following dialog box:

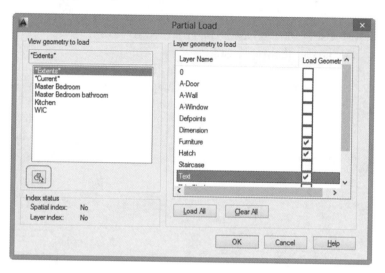

- As you can see this dialog box matches the Partial Open dialog box, with the exception of the small button at the lower-left portion of the dialog box. This button allows you to specify a window in the drawing to specify the extents of the objects to be loaded.
- Saving the partially loaded file means AutoCAD will keep these settings working until you change them. So, when you try to open it again, the following dialog box will be shown:

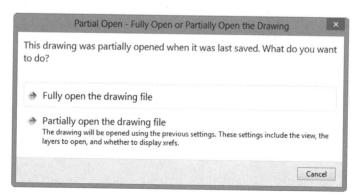

PRACTICE 13-6

 Using Partially Opened Files

1. Start AutoCAD 2014.
2. Using the Partial Open command, partially open **Practice 13-6.dwg** using Master Bedroom view and A-Door, A-Wall, and A-Window layers.
3. Using Partial Load, select layer Furniture.
4. Save and close the file.
5. Open it again using the normal Open command, and there will be a message asking you either to open it fully or partially; select partially.
6. Zoom out to see that all of the furniture is shown.
7. Using Partial Load again, select the Hatch layer and click the Pick a window button; select the area of the kitchen, then click OK.
8. Save and close the file.

NOTES:

CHAPTER REVIEW

1. The Content Explorer:
   a. Has a Google-like index that allows it to search and find the desired content quickly.
   b. Will add watched folders.
   c. Will search for a word in a drawing.
   d. All of the above.
2. The _____ command will work only on partially opened files.
3. Using the Select Similar command:
   a. You can select all circles.
   b. You can select all circles in the same layer.
   c. You can select all circles with the same color.
   d. All of the above.
4. Which one of the following statements is *not* true?
   a. You can isolate objects and isolate layers.
   b. A formula in a cell should start with an equals sign and includes cell addresses.
   c. You need to empty the layer in order to delete using the Delete in Layers panel.
   d. You can hide objects rather than hide layers.
5. You can insert the area of a polyline as a field without a background.
   a. True
   b. False
6. _____ will show a dialog box listing all layers in the current file. To show the contents of a layer, click its name in the list.

CHAPTER REVIEW ANSWERS

1. d
3. d
5. a

14

USING BLOCK
TOOLS AND
BLOCK EDITING

In This Chapter

◇ How to use Automatic Scaling
◇ How to use Design Center
◇ How to use and customize Tool Palette
◇ How to edit a block

14.1 AUTOMATIC SCALING

- When we created the block we input the Block unit, so each unit we use in this block will represent a certain unit. In order for this step to be complete, you have to set the drawing file unit. Go to the **Application Menu**, select **Drawing Utilities**, and then select **Units**. You will see the following dialog box:

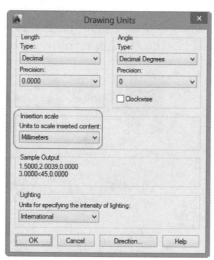

- Under **Units to scale inserted content** select the unit that will represent your drawing file unit. AutoCAD will convert the block unit to the drawing unit and will show it in the **Insert** dialog box under **Block unit**.

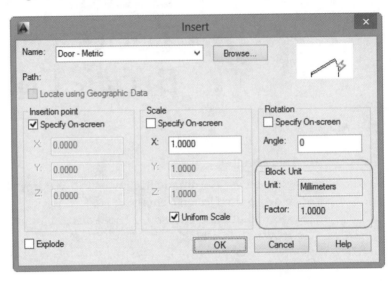

14.2 DESIGN CENTER

- Before AutoCAD 2000, there was no direct method for sharing blocks, layers, and other things. In AutoCAD 2000, **Design Center** was introduced as the ultimate solution to this problem. Using Design Center, you can share blocks, layers, dimension styles, text styles, table styles, etc.
- To issue the command go to the **Insert** tab, locate the **Content** panel, and then click the **Design Center** button:

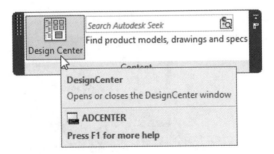

- You will see the following palette:

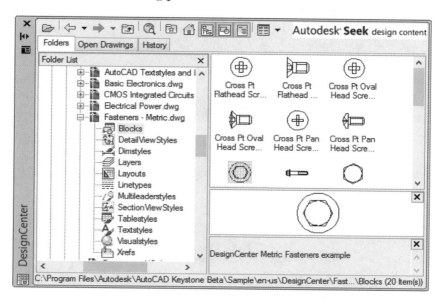

- The left pane of the Design Center is like My Computer in Windows, and it contains all of your drives, folders, and files. This allows you to locate the file that contains the needed blocks, layers, etc. When you locate the file, click at the plus sign at the left of the file name and a list will open. See the following illustration:

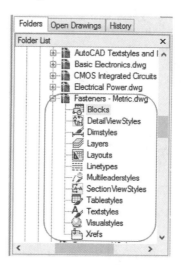

- Select the desired content. If you select Blocks, in the right pane you will see the shapes and the names. There are three methods to copy blocks to your current drawing using Design Center:
  - Drag-and-drop using the left mouse button to insert the block in the current file.
  - Drag-and-drop using the right mouse button. When you release it, you will see the following menu. Select the **Insert Block** option to make the Insert dialog box pop-up:

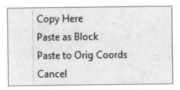

- Double-click, and the Insert dialog box will open.

PRACTICE 14-1

   Using Design Center

1. Start AutoCAD 2014.
2. Open **Practice 14-1.dwg**.
3. Make layer Furniture the current layer.
4. Start Design Center.
5. Locate your AutoCAD 2014 folder, then go to the following path: \Sample\en-us\Design Center (en-us folder is assumed for the English language AutoCAD, this folder may change depending on the language chosen)
6. Locate the file *Home-Space Planner.dwg*, and choose Blocks to insert the furniture shown below.
7. Make layer Toilet the current layer.

8. Locate the file *House Designer.dwg*, and choose Blocks to insert the toilet furniture as shown below.

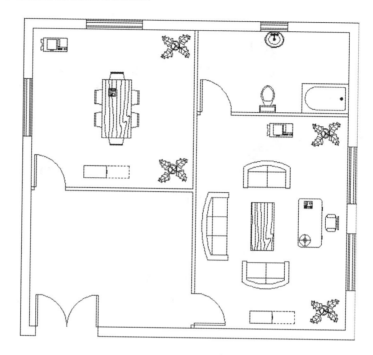

9. Save and close the file.

14.3 TOOL PALETTES

- Using Design Center to share data between files is a huge help, but there are still some lingering issues. You have to remember the current layer, the scale, the rotation angle, and search for the desired file each time you need to copy something from it. In 2004, AutoCAD introduced us to Tool Palettes, which take blocks to the next level.
- Tool Palettes allow you to store any type of object and then retrieve it in any opened file. You can also control the object's properties, so next time you drag it into your file, you will not have to worry about layers, scale, or rotation angles. You can keep several copies of the same block, and each can hold different properties.

- To start the command, go to the **View** tab, locate the **Palettes** panel, and then click the **Tool Palettes** button:

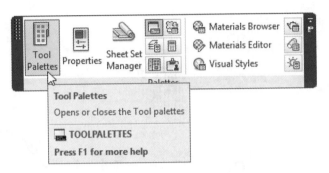

- You will see something like the following:

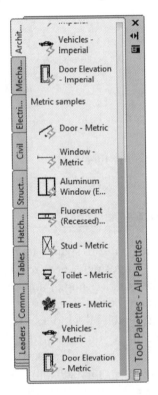

- AutoCAD includes a few Tool Palettes, but you can also create your own.

14.3.1 How to Create a Tool Palette from Scratch

- This method will create an empty palette, and you can fill it using different methods. To do this right-click over the name of any existing tool palette, and you will see the following menu:

- Choose the **New Palette** option, and a new empty palette will be created, allowing you to name it. Type in the name of the new palette as shown below:

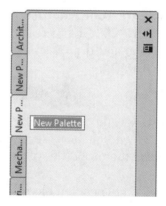

14.3.2 How to Fill the New Palette with Content

- You can fill the new empty palette with content using the drag-and-drop method from different drawings to the palette. For example, say you opened one of your colleague's drawing files and you discovered that she created several new blocks. Simply click the block, avoiding the grips, hold, and drag it to the palette. You can do the same thing with other objects (lines, arcs, circle, polylines, hatches, tables, dimensions, etc.).

14.3.3 How to Create a Palette from Design Center Blocks

- AutoCAD allows you to create a palette from all blocks in a file using Design Center. In order for this to work, make sure that both Design Center and Tool Palettes are displayed on the screen. Go to your desired file in the left pane of the Design Center, right-click, and you will see something like the following:

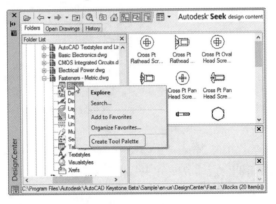

- Select the **Create Tool Palette** option and a new tool palette with the same name as the file will be created containing all the blocks.

NOTE ► - You can also drag-and-drop any block in any file from the Design Center to any tool palette.

14.3.4 How to Customize Tool Palette Properties

- You can create several copies of blocks and hatches in tool palettes using Copy/Paste. Once you have several copies of the same block/hatch, you can change the properties of these copies according to your needs. Follow these steps:
  - Right-click on the copied block/hatch, and you will see the following menu:

- Choose the **Properties** option, and you will see something like the following dialog box:

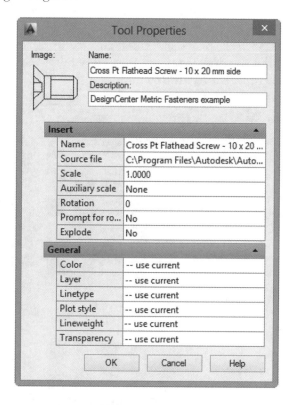

- Enter a new **Name** and **Description**. The Properties of the block/hatch are divided into two categories:
  - **Insert** type of properties
  - **General** type of properties
- By default, the **General** properties are all **use current**. For a block or a hatch you can specify that whenever you drag from the tool palette the block or hatch will reside in a certain layer (regardless of the current layer) and will have a certain color, linetype, lineweight, etc.
- With this feature, you can create your own tool palettes, which hold all the needed blocks and hatches, customized according to company standards, and designing will become a simple drag-and-drop process. If we know that 30–40% of drawings are blocks and 10–20% are hatches, if you use Tool Palettes effectively, your drawing time will be reduced significantly.

14.4 HATCHES AND TOOL PALETTES

- Just like blocks, you can store hatches in tool palettes. Storing hatches in tool palettes is not as important as storing hatches with altered properties, such as layer, scale, color, etc., to ensure simple and fast hatch insertion in the drawing. Go to the **View** tab, locate the **Palettes** panel, and select the **Tool Palettes** button:

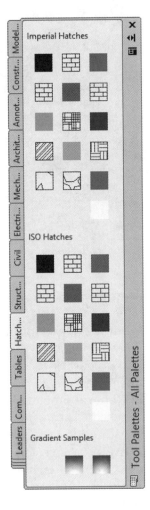

- Drag-and-drop from the drawing and to the drawing. Inside the tool palette right-click and choose the **Properties** option to change the properties of the saved hatch.

PRACTICE 14-2

Using Tool Palettes

1. Start AutoCAD 2014.
2. Open **Practice 14-2.dwg**.
3. Open Design Center, and locate /Sample/Design Center/Fasteners-US.dwg.
4. Locate the block underneath it, right-click, and then select the Create Tool Palette option.
5. If Tool Palettes is not displayed, it will be displayed with a new palette called Fasteners-US. Close Design Center.
6. Make sure that the current layer is 0.
7. Using the newly created tool palette, locate Hex Bolt ½ in. -side, right-click, and select Properties. Change the layer to Bolts.
8. Create another copy of it, and make the Rotation angle = 270.
9. Using drag-and-drop, drag the two blocks to the proper places as shown below.
10. Using the newly created tool palette, locate Square nut ½ in. -top, and set its layer to be Nut.
11. Right-click the copy, select Properties, and change the layer to Nut and the scale to 1.5.
12. Using drag-and-drop, drag the new block to the proper places as shown below.
13. Erase the lines to get the following shape:

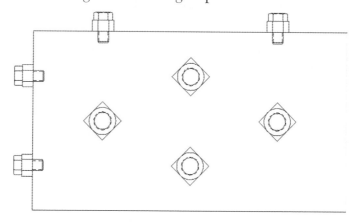

14. Close Tool Palettes. Save and close the file.

14.5 CUSTOMIZING TOOL PALETTES

- You can customize the look of tool palettes, i.e., set the transparency, view options, and so on. All of these functions will be available at the right-click menu.

14.5.1 Allow Docking

- You can dock the tool palette to the right or left of the screen to make it permanent. The default mode is floating. To dock it, make sure the tool palette is shown, right-click any empty space within the tool palette (avoid icons), and you will see the following menu. Select Allow Docking (if you see (✓) at its left, this means it is on):

- Now you can drag the tool palette to the right or left of the screen.

14.5.2 Transparency

- You can also set the transparency value of the tool palette. The default value is 100% opacity. To do this, make sure the tool palette is shown, right-click any empty space within the tool palette (avoid icons), and you will see the following menu. Select the Transparency option:

- You will see the following dialog box:

- You can change the value of the transparency for the palette (use the slider under General), and for Rollover (when your mouse is over the palette). The value of the transparency of Rollover should be always equal to or more than the general value.
- Also, select whether these settings affect the current palette or all palettes, and whether to disable all window transparency or not.

14.5.3 View Options

- You can set how the icons will appear inside the palette. To do this, make sure that tool palette is shown, right-click any empty space within the tool palette (avoid icons), and you will see the following menu. Select the View Options option:

- You will see the following dialog box:

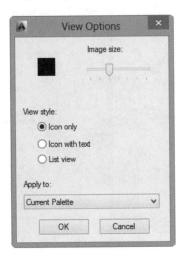

- Using this dialog box, you can:
  - Set the size of the image
  - Set the view style: Icon only, Icon with text, or List view
  - Whether to apply these changes to the current palette or all palettes

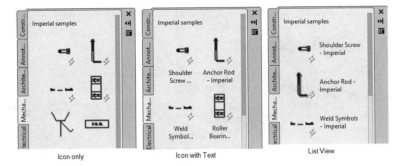

14.5.4 Add Text and Add Separator

- You can also add text to create internal groupings for each palette, or add lines to use as separators. To do this, make sure the tool palette is shown, right-click any empty space within the tool palette (avoid icons), and you will see the following menu. Select the Add Text and Add Separator options:

- This is what you will get:

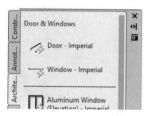

14.5.5 New/Delete/Rename Palette

- These three commands allow you to create a new palette, delete an existing palette, or rename an existing palette. To do this, make sure the tool palette is shown, right-click any empty space within the tool palette (avoid icons), and you will see the following menu. Select the New Palette, Delete Palette, and Rename Palette options:

14.5.6 Customize Palettes

- You can create palette groups to organize them. You can then set one of the groups to be the current palette group. To do this, make sure the tool palette is shown, right-click any empty space within the tool palette (avoid icons), and you will see the following menu. Select the Customize Palettes option:

- You will see the following dialog box:

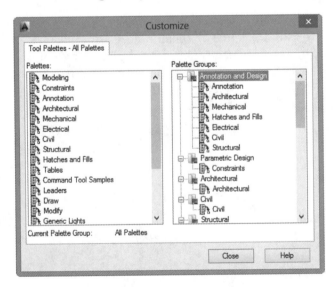

- At the right, you will see the predefined palette groups along with the palettes belonging to them. At the left, you will see the defined palettes. In order to create a new group, at the right click anywhere, and you will see the following menu:

- Select the New option, and type the name of the new group. You may move the group from any place to any place using drag-and-drop.
- To fill this group with palettes, use drag-and-drop again from left part to right part.
- To make this group the current group, right-click it, then select the Set Current option:

- Using this menu you can Rename a group, Delete a group, Export one group, Export all groups, and finally Import groups. The palette group file extension is *.XPG.

PRACTICE 14-3

Customizing Tool Palettes

1. Start AutoCAD 2014.
2. Open **Practice 14-3.dwg**.
3. Start the Tool Palette command.

4. Using the tool palette created from Practice 14-2 rearrange the tools in Fasteners to see all nuts together, all screws together, and all bolts together.
5. Add separators between the three types, and add text as titles for each type.
6. Change the transparency of the tool palette to be 50% for general.
7. Using the View Options, change the size of the tool to be less than maximum with one degree, showing Icons with text.
8. Using Customize Palettes create a new group and call it My Group.
9. Drag-and-drop Fasteners – US to it.
10. Make My Group the current group (you can see that this group contains only one tool palette).
11. Export My Group to be **My Group.xpg** (you can go to another computer and try to import the same).
12. Save and close the file.

14.6 EDITING BLOCKS

- You can edit the original block in Block Editor, which is used to create the dynamic features of a block (this is an advanced feature of Auto-CAD). Block Editor allows you to redefine the block by adding/removing/modifying the existing objects.
- To issue the command, go to the **Insert** tab, locate the **Block Definition** panel, and then click the **Block Editor** button:

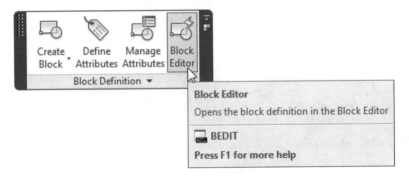

- You will see the following dialog box:

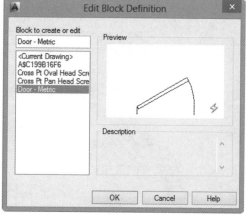

- Select the name of the desired block to edit and then click **OK** to start editing. Another easy way is to double-click one of the blocks.
- When the Block Editor opens, you will see several things take place: the background color will be different, a new context tab called "Block Editor" will appear, and several new panels will appear as well. Ignore all of this, and start adding, removing, or modifying the objects of your block. Once you are done, click the **Save Block** button on the **Open/Save** panel (you can find it at the left), and then click the **Close Block Editor** button in the **Close** panel to end the command.

PRACTICE 14-4

Editing Blocks

1. Start AutoCAD 2014.
2. Open **Practice 14-4.dwg**.
3. Double-click one of the Single Door blocks.
4. Change the arc linetype to Dashed2.
5. Save the block with the new changes.
6. What happened to the other Single Door blocks? _____

7.   You should have the following shape:

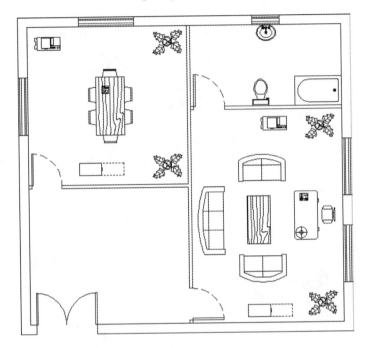

8.   Save and close the file.

NOTES:

CHAPTER REVIEW

1. Tool Palettes can store commands such as line, polyline, etc.
   a. True
   b. False
2. _____ is the tool that allows you to share blocks, layers, etc.
   a. Insert Center
   b. Design Office
   c. Design Center
   d. The Internet
3. Using Tool Palettes you can customize all tools by changing their properties.
   a. True
   b. False
4. Double-clicking a block will show the_____ dialog box.
5. If both Design Center and Tool Palettes are open, you can drag any block from Design Center to a tool palette.
   a. True
   b. False
6. Using _____ of the right-click menu will give you Icon only, Icon with text, and List view:
   a. Transparency
   b. View Options
   c. Auto Hide
   d. Allow Docking

CHAPTER REVIEW ANSWERS

1. a
3. a
5. a

15

CREATING TEXT & TABLE STYLES AND FORMULAS IN TABLES

In This Chapter

◇ How to create a Text Style
◇ How to create a Table Style? Then using the Table command to insert Table
◇ In Tables, how to use Formulas
◇ In Tables, how to utilize the cell functions

15.1 CREATING TEXT AND TABLES

- Inserting text in AutoCAD involves two simple steps:
  - Creating a text style (normally will be created once) that includes the size and the shape of the text
  - Inserting text using either Single Line Text or Multiline Text
- Creating text styles can be a tedious and lengthy job and is usually the job of CAD managers, since standardization is important. Table styles are also a part of the process of standardization and include:
  - Creating a table style
  - Inserting and filling a table
- Text and table styles and all other styles should be a part of the document template that holds company standards. If you don't have a template, you can still share styles using Design Center (as discussed in Chapter 6).

15.2 HOW TO CREATE A TEXT STYLE

- The first step in adding text in AutoCAD is to create a text style, which is where you define the characteristics of your text. To start the **Text Style** command, go to the **Annotate** tab, locate the **Text** panel, and click the small arrow at the lower right:

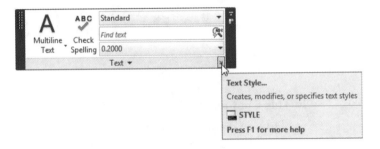

- The following dialog box will appear:

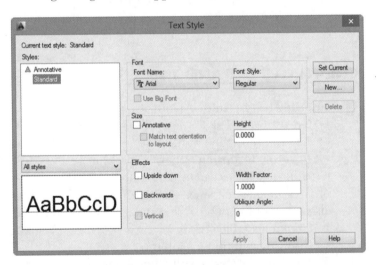

- There are two pre-defined text styles; one is called **Standard** and the other is called **Annotative**. They are almost the same, except the latter uses the Annotative feature (which will be discussed in Chapter 9). Both styles use the Arial font. Professional users usually create their own text styles. To create a new text style, click the **New** button, and you will see the following dialog box:

- Enter the name of the new text style and click **OK**. The first thing you will do is select the **Font**. There are two types of fonts you can use in AutoCAD:
  - Shape File Fonts (*.shx) the very old type of fonts (out-of-date)
  - True Type fonts (*.ttf)
- See the illustration below and notice how .ttf files are more accurate and look better:

True type font   Shape file font

- Next, you will select the **Font Style** (that is, if you selected a true type font). You will choose one of the following:
  - Regular
  - Bold
  - Bold Italic
  - Italic
- Keep **Annotative** off for now (it will be discussed later). Next, you have to specify the **Height** of the text, which is the height of the capital letters (lowercase letters will be two-thirds of the height). See the following illustration:

Text height is the Capital letter height

- You have two options when setting text height:
  - You can set the height to 0 (zero), which means the height will be variable (you will have to input the value each time you use this style).
  - You can set the height to a value greater than 0 (zero), which means the height is fixed.
- Finally, you will set the effects. You have five of them:
  - Upside down, to write text upside down.
  - Backwards, to write text from right to left.
  - Vertical, to write from top-to-bottom (good for Chinese words, but only for Shape files).
  - Width Factor, to set the relationship between the width of the letter and its length. If the value>1.0, then the text will be wide. If the value is <1.0, the text will be long.
  - Oblique Angle, to set the angle to italicize either to the right (positive value) or to the left (negative value).
- When you are done, click the **Apply** button, and then **Close**. You can show **All styles** or show **Styles in use**. See the pop-up list at the left as shown below:

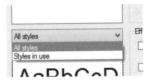

PRACTICE 15-1

  Creating a Text Style and Inserting Single Line Text

1. Start AutoCAD 2014.
2. Open **Practice 15-1.dwg**.
3. Create a text style with the following specs:
   a. Name = Room Names
   b. Font = Tahoma
   c. Annotative = off
   d. Height = 0.3
   e. Leave the rest as default values

4. Make the Room Names text style current.
5. Make layer Text current.
6. Enter the room names as shown below.
7. Make Standard text style current (with this text style the height = 0, so you should set it every time you want to use this style).
8. Make layer Centerlines current.
9. Zoom to the upper-left centerline and notice that the letter A is missing.
10. Start Single Line Text, right-click, and select the Justify option. From the list choose MC (Middle Center), select the center of the circle as the Start point, set the Height set to 0.25 and the Rotation angle = 0, then type A, and press [Enter] twice.
11. You should have the following:

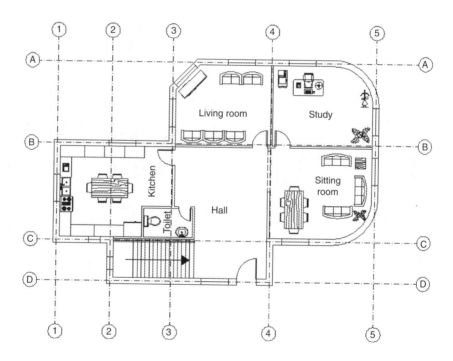

12. Save and close the file.

15.3 CREATING A TABLE STYLE

- In order to create a professional table, you should create a table style, which holds the features of the table and specifies the Title, Header, and the Data rows. Using this style you can insert as many tables as you wish. Table styles can be shared using Design Center.
- To issue the command, go to the **Annotate** tab, locate the **Tables** panel, and then select the **Table Style** button:

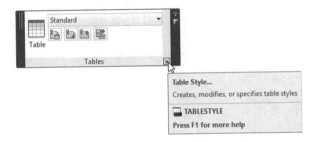

- The following dialog box will be displayed:

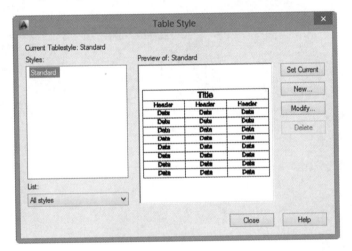

- Just like for text styles, there is a pre-defined table style called Standard. Click the **New** button to create a new table style, and you will see the following dialog box:

- Enter the name of the new table style and click **Continue**.
- You will see the following dialog box:

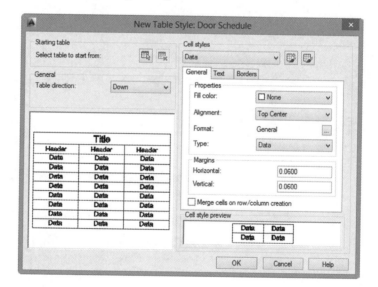

- AutoCAD allows you to start a new table style based on an existing table in your current drawing; click the button shown here:

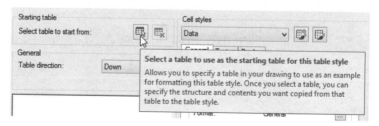

- If you want a new style, you need to specify the characteristics of your table by selecting the **Table direction**, **Down** or **Up**; see the following illustration:

- AutoCAD creates three parts for any table: Title, Header, and Data. Using the Table Style command you can set these three parts by selecting the desired part, then using the **General** tab, **Text** tab, and **Border** tab to create your settings:

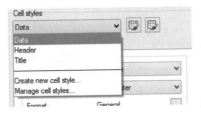

15.3.1 General Tab

- The General tab is shown below:

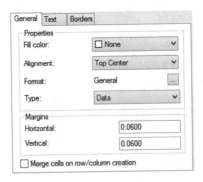

- Here, you can do the following:
  - Change the Fill Color of the cells (by default it is None).

- Change the Alignment to set up the text related to the cell borders. For example, if you choose Top Left, the text will reside in the top-left part of the cell. See the following illustration:

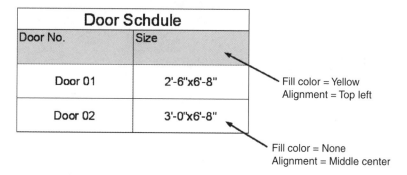

- Change the Format. You will see the following dialog box, which allows you to set the format of the cell, whether it is currency, percentage, date, etc.:

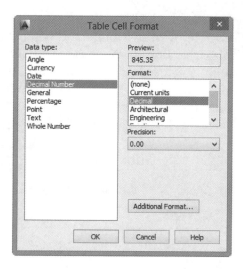

- Change the Type of the contents of the cell to Data or Label. This part is very important, since some of the cells may hold numbers, but

these numbers shouldn't be included in a mathematical formula, so the type of data will be Label and not Data.
- Under Margins, control the Horizontal and Vertical distances around the Data relative to the borders.

15.3.2 Text Tab

▪ The Text tab is shown below:

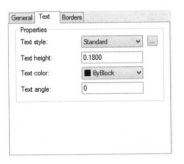

▪ Here, you can do the following:
  - Change the Text style of the text filling the cells.
  - Change the Text height of the text filling the cells (keeping in mind that if the text style has a height > 0.0, the number here is meaningless).
  - Change the Text color.
  - Change the Text angle; see the following illustration:

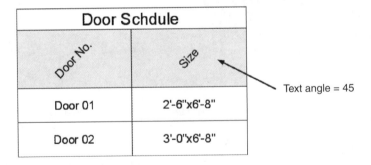

15.3.3 Borders Tab

- The Borders tab is shown below:

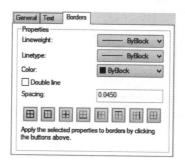

- Here, you can:
  - Change the Lineweight, Linetype, and Color of the border lines.
  - Change the border to a Double line instead of a Single line (default value), and also choose the Spacing between the lines.
  - Change whether you want lines representing column separators and row separators.

15.4 INSERTING A TABLE IN A DRAWING

- This command allows you to insert a table in the current drawing. To issue this command, go to **Annotate** tab, locate the **Tables** panel, and then select the **Table** button:

- You will see the following dialog box:

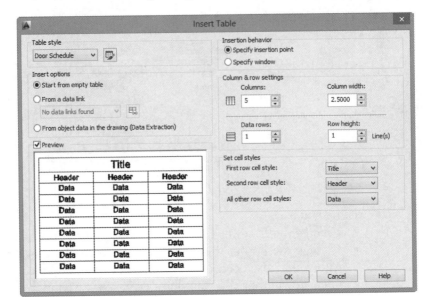

- The first step is to select the desired pre-defined **Table style** name from the available list. If you haven't created a table style yet, click the small button beside the list to define it now.
- Select the proper **Insert options**. To insert the table in your drawing, choose one of the following three choices:
  - Start from empty table.
  - From a data link – this is an advanced feature of AutoCAD.
  - Start from object data in the drawing (Data Extraction) – this is also an advanced feature of AutoCAD.
- Most of the time you will use the first option, **Start from empty table**. Now select the **Insertion behavior**; there are two available choices:
  - Specify insertion point.
  - Specify window.

15.4.1 Specify Insertion Point Option

- The insertion point here is the upper-left corner of the table. Use the following data for the rest of the table:
  - Columns (the number of columns)
  - Column width

- Data rows (the number of rows, but without Title and Heads)
- Row height (in lines)

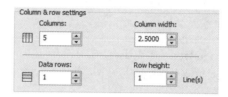

- Click **OK**, and AutoCAD will show the following prompt:

```
Specify insertion point:
```

- Specify the location of the table and start filling the cells. You can use the arrows on the keyboard or the [Tab] key to jump from one cell to another. Use [Shift] + [Tab] to go backwards.

15.4.2 Specify Window Option

- Using this option you will specify a window, which means you will give AutoCAD the total length and the total width of the table. In order to fulfill the rest of the information, input the following:
  - The number of Columns, and AutoCAD will calculate the column width. Or input the Column width, and AutoCAD will calculate the number of columns.
  - The number of Data rows, and AutoCAD will calculate the row height. Or input the Row height, and AutoCAD will calculate the number of rows:

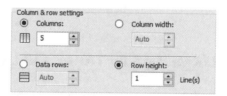

- Click **OK**, and the following prompts will appear:

```
Specify first corner:
Specify second corner:
```

- Specify the two opposite corners of the table, then start inputting the cell contents using the arrows, [Tab], and [Shift] + [Tab].

PRACTICE 15-2

Creating a Table Style and Inserting a Table in a Drawing

1. Start AutoCAD 2014.
2. Open **Practice 15-2.dwg**.
3. Create a new table style with the following settings:
   a. Name = Door Schedule
   b. Title Text Style = Notes
   c. Header Fill Color = Yellow
   d. Header Text Style = Notes
   e. Data Alignment = Middle Left
   f. Data Horizontal Margin = 10
   g. Data Text Style = Notes
   h. Data Text Color = Blue
4. Create a table using the above table style and the frame drawn using the Window option to insert the table:

Door Schedule	
Door No.	Size
Door 01	2'-6" x 6'-8"
Door 02	3'-0" x 6'-8"

5. After finishing the table input, erase the frame.
6. Save and close the file.

15.5 USING FORMULAS IN TABLE CELLS

- In this section, we will discuss how to create formulas in AutoCAD table cells, along with more functions like merging cells, using premade functions, and others. If you know how to use Excel, you can skip this part; if not read it because you will need this information in the following sections.

- We call the intersection of a column and row in a table a **Cell**.
- The cell address comes from the column letter and row number. For example, B15 means column B and row 15.
- To make sure AutoCAD knows you will be writing a formula, start with an equals sign. Along with the cell addresses, you can use arithmetic functions like add, subtract, multiply, and divide. For example, =(B3/2)+A9.
- AutoCAD provides some premade functions such as SUM, COUNT, and AVERAGE.
- To allow users to copy formulas from one cell to another, AutoCAD opted to make the cell address relative. For example, let's say we have a formula in cell C3, =A3+B3. If we copy it to C4, what is the result? AutoCAD will change the formula to =A4+B4. AutoCAD provides a tool called *Autofill grip* to copy formulas (or any cell content). You can click and drag it upwards or downwards:

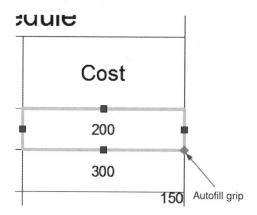

- Adding a dollar sign $ will attach the formula either to a column or to a row, or to both. For example, $F12 means column F is the column to use while copying formulas, with variable rows. While if we type F$12, row 12 is the row to use while copying, with variable columns. F12 is a fixed address.

15.6 USING TABLE CELL FUNCTIONS

- There are three different modes for table editing:
  - If you click the outer frame, you will edit the whole table by adjusting the total width/height for the whole table or columns and rows, just like the following:

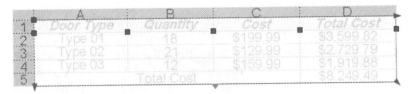

  - If you click a cell or group of cells (to select a group of cells, click and drag) a new context tab called Table Cell will be added to control the cells.
  - If you double-click inside a cell, you will be able to input data, or edit existing data, and the Text Editor context tab will appear.
- In the following, we will discuss the Table Cell context tab and its contents.

15.6.1 Using the Rows Panel

- You will use the following panel:

- You can do all or any of the following:
  - Insert a new row **Above** the selected cell(s)
  - Insert a new row **Below** the selected cell(s)
  - **Delete** the selected **Row(s)**

15.6.2 Using the Columns Panel

- You will use the following panel:

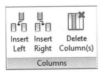

- You can do all or any of the following:
  - **Insert** a new column to the **Left** of the selected cell(s)
  - **Insert** a new column to the **Right** of the selected cell(s)
  - **Delete** the selected **Column(s)**

15.6.3 Using the Merge Panel

- You will use the following panel:

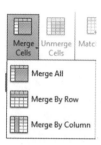

- You can do all or any of the following:
  - Merge all selected cells in a single cell
  - Merge selected cells using rows
  - Merge selected cells using columns
  - If you have cells containing different types of data (numbers, text, dates, etc.) which one will be used? The answer is the content of the first cell. You will see the following message:

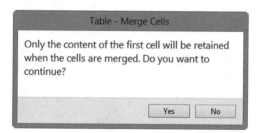

  - Unmerge cells to get them back to their previous condition before merging.

15.6.4 Using the Cell Styles Panel

- ■ You will use the following panel:

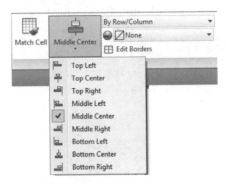

- ■ You can do all or any of the following:
  - • **Match cell** to match the properties of a selected cell with other cells
  - • Select the **Alignment** using one of the nine available alignments for the content relative to the cell borders
  - • Select one of the existing cell styles for the selected cells:

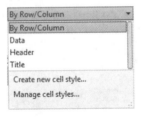

  - • Select a color to be the background for the selected cells:

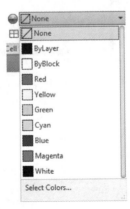

- Click the Edit Borders button, and the following dialog box will appear:

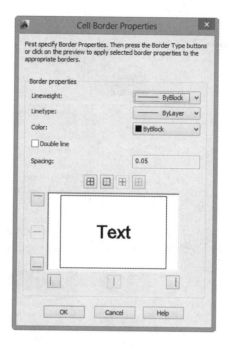

- You can control all properties of the border such as Lineweight, Linetype, Color, Double line, and Spacing, and whether or not to show border lines in different locations.

15.6.5 Using the Cell Format Panel

- You will use the following panel:

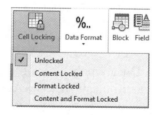

- You can do all or any of the following:
  - Select whether to lock the contents or format, or both, of the selected cells. You can unlock the locked cells as well.

- Select the Data Format of the selected cells choosing one of the following:

- You can also customize the data format. Select the **Custom Table Cell Format** option, and you will see the following dialog box:

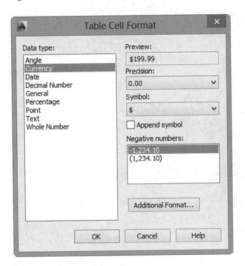

- Select the **Data** type you want to change, and then make the necessary changes. You can click the **Additional Format** button to see the following dialog box and to specify what to use for the Decimal and

Thousands separator and whether or not to suppress Leading and Trailing zeros:

15.6.6 Using the Insert Panel

- You will use the following panel:

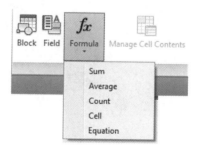

- You can do all or any of the following:
  - Click the Block button, and the following dialog box will appear:

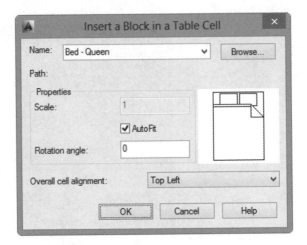

- Select the block name or click the **Browse** button to use any block or file, then specify the block's **Scale** or **AutoFit** and the **Rotation angle**. Lastly, specify the Overall cell alignment of the block reference to the cell border.
- Click the Field button to insert a field in the cell.
- Click the Formula button to insert a premade formula.

15.6.7 Using the Data Panel

- You will use the following panel:

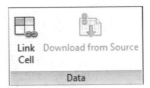

- You can do all or any of the following:
  - Click the **Link Cell** button, and the following dialog box will appear:

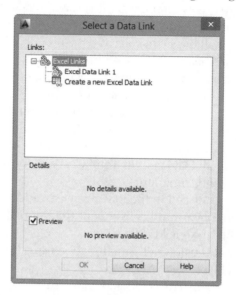

  - This is the same dialog box that we dealt with while linking data with Excel. Create the link that you want and then use the **Download from Source** button to bring in the data from this link.

PRACTICE 15-3

 Formulas and Table Cell Functions

1. Start AutoCAD 2014.
2. Open **Practice 15-3.dwg**.
3. Insert the block in the first column using the second column as your guide (for the first two blocks use AutoFit and for the rest use Scale = 0.01).

4. Select all Data cells and make them Middle Center.
5. Select one of the Cost column cells.
6. Insert a new column to the left and name it Quantity.
7. Input the following values in it from top to bottom: 4, 1, 1, 6, 4.
8. Select all the Data cells of the Cost column.
9. Select Custom Table Cell Format (from Cell Format panel) and change the currency format to show a number like the following: $999.99.
10. Select one of the Cost column cells.
11. Insert a new column to its right and call it Total Cost.
12. Input a formula: Quantity * Cost.
13. Using AutoFill grip copy it to the other cells.
14. Select all Data cells of Total Cost and make sure to add a comma as the thousands separator.
15. Select any cell in the lower row and add a new row below it.
16. Merge by row all cells except the rightmost cell and type Total Cost in it and make it Middle Right.
17. Add at the new cell Sum function using the range of cells containing the Total Cost.
18. Select the Title cell, and make sure that the border lines at the top, right, and left are removed.
19. Set the background color of the Header to be Cyan.
20. Save and close the file.

NOTES:

CHAPTER REVIEW

1. In _____, you will define the background color of the table cells.
2. There are two methods to insert a table in a drawing.
   a. True
   b. False
3. By default the first row cell style is _____.
4. Height in a text style is for:
   a. Small letters
   b. Capital letters and small letters
   c. Capital letters only
   d. Everything above the baseline
5. When inserting a table using the Window option, specifying the number of columns is enough; you don't need to specify the column width.
   a. True
   b. False
6. In a table style, you can specify different text styles for Header and Title.
   a. True
   b. False

CHAPTER REVIEW ANSWERS

1. Table Style
3. Title
5. a

16

DIMENSION & MULTILEADER STYLES

In This Chapter

◇ How to create a dimension style
◇ More advanced dimension commands
◇ How to create a multileader style
◇ How to insert a multileader in a drawing

16.1 WHAT IS DIMENSIONING?

- Dimensioning in AutoCAD is just like using text and tables. You should create your dimension style first, and then use it to insert dimensions. Dimension styles control the overall outcome of the dimension block generated by the different types of dimension commands.

- To insert a dimension, depending on the type of the dimension, you should specify points, or select objects, and then a dimension block will be added to the drawing. For example, in order to add a linear dimension you will select two points representing the distance to be measured, and a third point will be the location of the dimension block. See the illustration below:

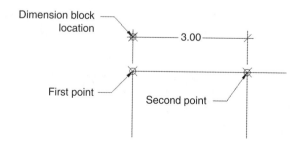

- The generated block consists of three things:
  - Dimension line
  - Extension lines
  - Dimension text
- See the following illustration:

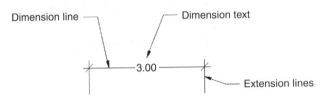

16.2 HOW TO CREATE A NEW DIMENSION STYLE

- This command allows you to create a new dimension style, or modify an existing one. To issue this command, go to the **Annotate** tab, locate the **Dimensions** panel, and then select **Dimension Style** button:

- You will see the following dialog box:

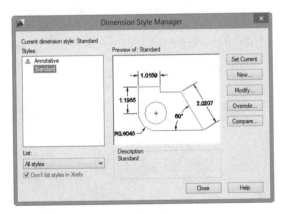

- As you can see there are two pre-defined dimension styles: Standard and Annotative. Click the **New** button to create a new style. You will see the following dialog box:

- Enter the name of the new style. Under **Start With**, select the existing style that you will use as your starting point. Leave the Annotative checkbox off (we will discuss it in Chapter 9). Make sure that Use for All dimensions is selected (we will cover it at the end of our discussion), and then click the **Continue** button.

16.3 DIMENSION STYLES: LINES TAB

- As a rule-of-thumb, and while we are discussing the different dimension style tabs, we will leave Color, Linetype, and Lineweight at their default settings because we want these things to be controlled by the layer rather than the individual dimension block.
- The first tab in the dimension style dialog box is Lines, and it allows you to control the dimension lines and extension lines:

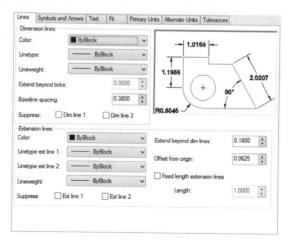

- Under **Dimension lines** change all or any of the following settings:
  - Control **Extended beyond ticks**, which is as illustrated below (this option works only when **Arrowhead** is **Architectural tick** or **Oblique**):

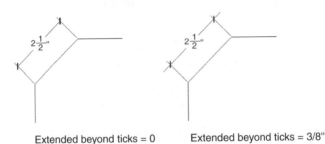

Extended beyond ticks = 0   Extended beyond ticks = 3/8"

  - As we will see in this chapter when you add a baseline dimension you will not control the spacing between a dimension and another, you will control the **Baseline spacing**, as illustrated below:

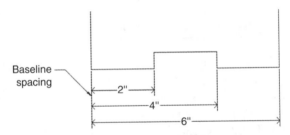

- From now on when we say First, we mean the nearest point to the first point chosen, and Second is the nearest point to the second point chosen.
  - Choose whether to **Suppress Dim line 1, Dim line 2**, or leave them as is. See the illustration below:

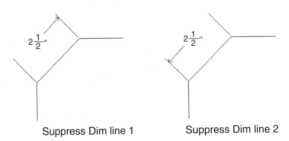

Suppress Dim line 1   Suppress Dim line 2

- Under **Extension lines** change all or any of the following settings:
  - Choose whether to **Suppress Ext line 1, Ext line 2,** or leave them as is. See the illustration below:

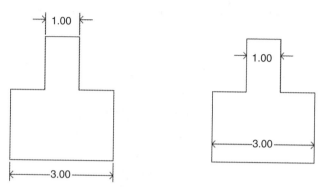

Both Extension lines are displayed   Suppress both Extension lines

  - Input **Extend beyond dim lines** and **Offset from origin**. See the illustration below:

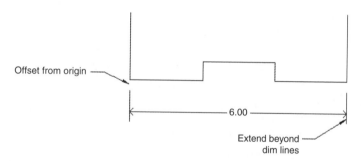

  - Choose whether to fix the length of the extension lines or not, and if yes, what the length will be. See the following example:

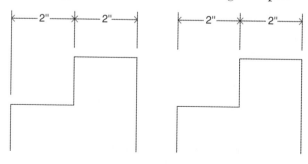

Fixed length = Off   Fixed length = On

16.4 DIMENSION STYLES: SYMBOLS AND ARROWS TAB

- This tab allows you to control the arrowheads and related features. The following shows the **Symbols and Arrows** tab:

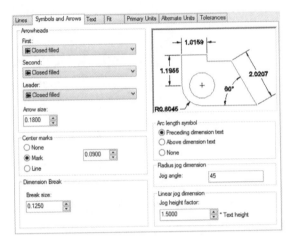

- Under **Arrowheads**, change all or any of the following:
  - The shape of the **First** arrowhead. When you set the shape of the first arrowhead, the **Second** arrowhead will automatically change, but you can also change them manually.
  - The shape of the arrowhead to be used in the **Leader** (Radius and Diameter are not leaders).
  - The **Size** of the arrowhead.
- Under **Center marks**, choose whether to show or hide the center mark of arcs and circles as shown below, then set the **Size** of the center mark:

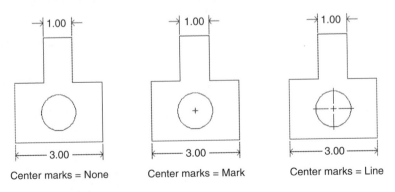

Center marks = None   Center marks = Mark   Center marks = Line

- Under **Dimension Break**, input the **Break size.** The break size is defined as the distance of the void left between two broken lines of a dimension. See the illustration below:

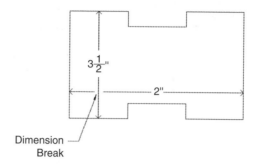

Dimension Break

- Under **Arc length symbol**, choose whether to show (as shown in the illustration below) or hide the arc length symbol:

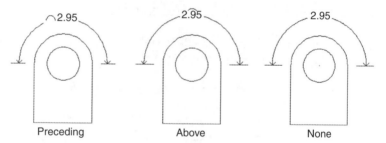

Preceding   Above   None

- Under **Radius dimension jog**, input the **Jog angle** as shown in the following illustration:

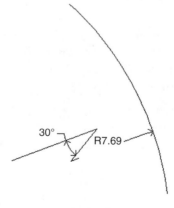

Jog angle

- Under **Linear Jog dimension**, input the **Jog height factor** as shown in the below. The Jog height factor is defined as a factor used to multiply the height of the text used in a dimension:

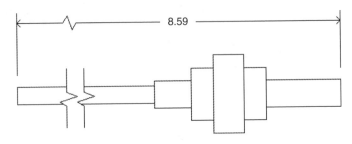

Jog height factor = 2.0

16.5 DIMENSION STYLES: TEXT TAB

- This tab allows you to control the text appearing in the dimension block. The following shows the **Text** tab:

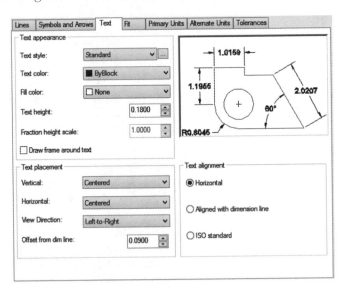

- Under **Text appearance** change all or any of the following:
  - Select the desired **Text style**, or create a new one.

- Specify the **Text color** and the **Fill color** (text background color):

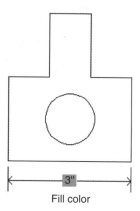

Fill color

- If your text style has a text height = 0 (zero), then input the **Text height**.
- Depending on the primary units (discussed in a moment) set **Fraction height scale**:

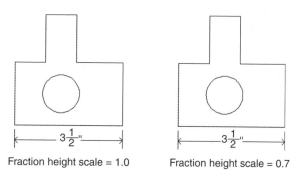

Fraction height scale = 1.0   Fraction height scale = 0.7

- Select whether your text will have a frame. See the illustration below:

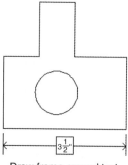

Draw frame around text

▪ Under **Text placement** change all or any of the following:
  • Choose the **Vertical** placement of your text related to the dimension line. There are five available choices: Centered, Above, Outside, JIS (Japan Industrial Standard), and Below. See the illustration below:

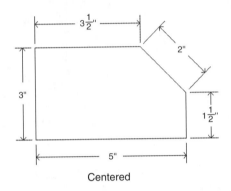

Centered

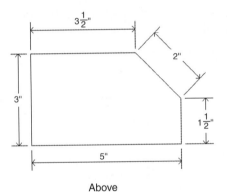

Above

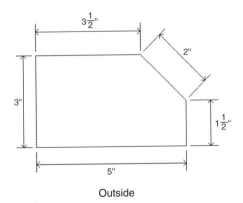

Outside

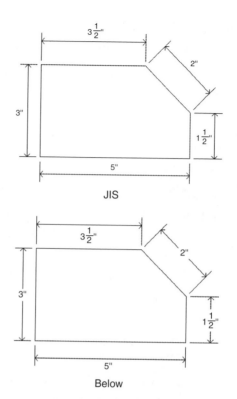

JIS

Below

- Choose the **Horizontal** placement. You have four choices:

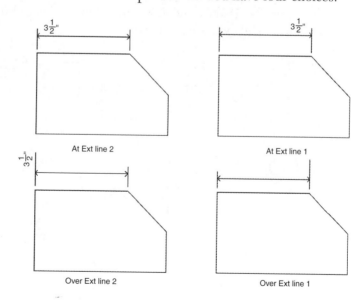

At Ext line 2

At Ext line 1

Over Ext line 2

Over Ext line 1

- Choose the **View Direction** of the dimension text: Left-to-Right or Right-to-Left. See the following illustration:

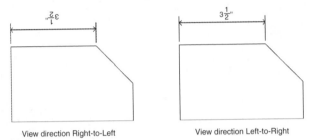

View direction Right-to-Left   View direction Left-to-Right

- Input the **Offset from dim line**, as shown below:

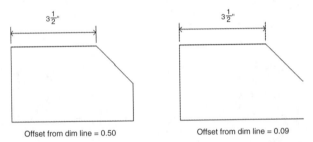

Offset from dim line = 0.50   Offset from dim line = 0.09

- Under **Text alignment**, control the alignment of the text related to the dimension line, whether always horizontal regardless of the alignment of the dimension line, or aligned with the dimension line. ISO will influence only the Radius and Diameter dimensions (all of the dimension types will be aligned except for those two):

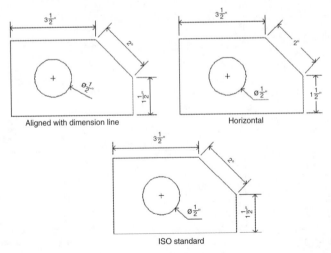

Aligned with dimension line   Horizontal

ISO standard

16.6 DIMENSION STYLES: FIT TAB

- This tab allows you to control the relationship between the dimension block components. The following shows the **Fit** tab:

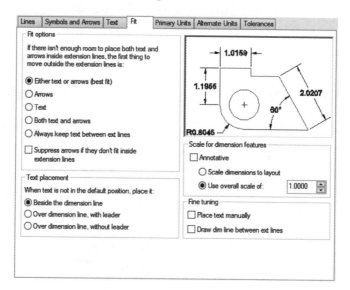

- Under **Fit options**, change all or any of the following:
  - If there is no room for the text and/or the arrowheads inside the extension lines, what do you want AutoCAD to do? Select the desired option.
  - If there is no room for the arrows to be inside the extension lines, do you want AutoCAD to suppress them?
- Under **Text placement**, when the text is not in the default position, select one of these options:

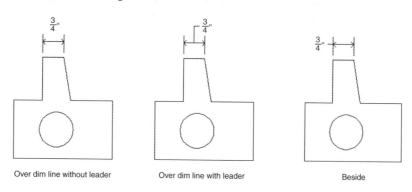

Over dim line without leader   Over dim line with leader   Beside

■ Under **Scale for dimension features**, you will control the size of the text (length, sizes, etc.). Will it be scaled automatically if it is Annotative? Or will it follow the viewport scale if it was input in the layout? If you want to input it in the Model space, you can set the scaling factor.

■ Under **Fine tuning**, select whether you want to place your text manually or leave it to AutoCAD. Also select whether to always force text inside extension lines.

16.7 DIMENSION STYLES: PRIMARY UNITS TAB

■ In this tab, you control everything related to the numbers that appear at the dimension block. The following shows the **Primary Units** tab:

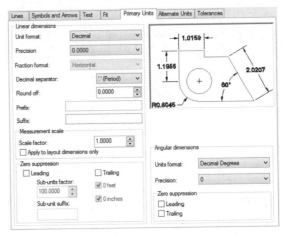

■ Under **Linear dimensions**, change all or any of the following settings:
  • Select the desired **Unit format**, then select its **Precision.**
  • If your selection was **Architectural** or **Fractional**, choose the desired **Fraction format**. You have three choices:

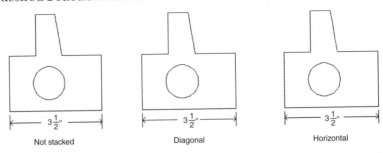

Not stacked   Diagonal   Horizontal

- If you select **Decimal**, choose the **Decimal Separator**. You have three choices: **Period**, **Comma**, and **Space.**
- Input the **Round off** number.
- Input the **Prefix** and/or the **Suffix**, as shown below:

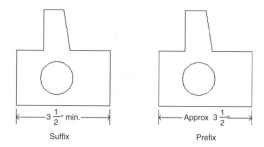

Suffix   Prefix

- Under **Measurement scale**, change all or any of the following:
  - By default AutoCAD will measure the distance between the two points specified by you (if it is linear) and input the text in the format set. But what if you want to show a different value than the measured value? Then you would input the **Scale factor**.
  - Choose whether this scale will affect only the dimension input in the layout. (We will discuss layouts in Chapter 9.)
- Under **Zero suppression**, choose whether to suppress the **Leading** and/or the **Trailing** zeros, as shown below:

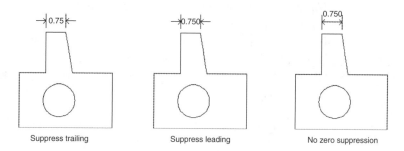

Suppress trailing   Suppress leading   No zero suppression

- If you have meters as your unit, and the measured value is less than one, this is a sub-unit. Select the sub-unit factor and the suffix for it (in this example it is cm).
- Under **Angular dimensions**, choose the **Units format** and the Precision. Control **Zero suppression** for angles as well.

16.8 DIMENSION STYLES: ALTERNATE UNITS TAB

- This tab allows you to show two numbers in the same dimension block; one showing the primary units and the other showing the alternate units. The following shows the **Alternate Units** tab:

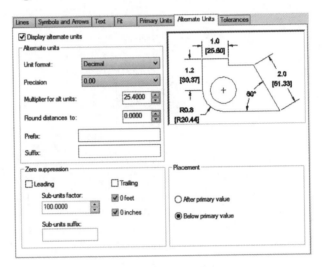

- Click on the **Display alternate units** option, then change the following:
  - Choose the Alternate **Unit format** and its **Precision**.
  - Input the **Multiplier** for all unit value.
  - Input the **Round** distance.
  - Input the **Prefix** and the **Suffix**.
  - Input the **Zero suppression** method.
  - Choose the method of displaying alternate units: **After primary value** or **Below primary value**. See below:

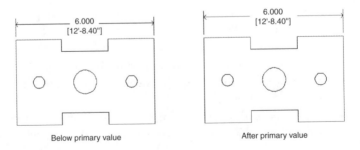

16.9 DIMENSION STYLES: TOLERANCES TAB

- This tab allows you to control whether or not to show tolerances and what method to use. The following shows the **Tolerances** tab:

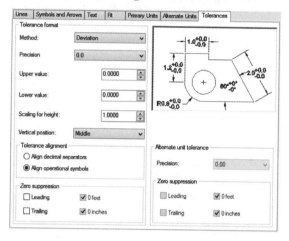

- There are four tolerance formats:
  - Symmetrical
  - Deviation
  - Limits
  - Basic
- The following is an illustration of each of the four choices:

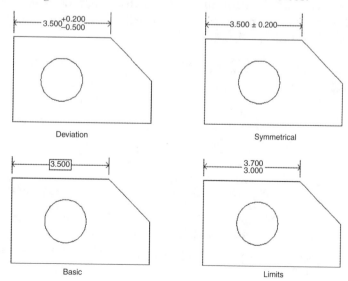

Deviation

Symmetrical

Basic

Limits

- Under **Tolerance format**, change all or any of the following:
  - Select the proper **Method** and then select its **Precision**.
  - Depending on the method specify the **Upper value** and **Lower value**.
  - Input **Scaling for height** for the tolerance values if desired.
  - Choose the **Bottom**, **Middle**, or **Top** vertical position for the dimension text with reference to the tolerance values. See the following illustration:

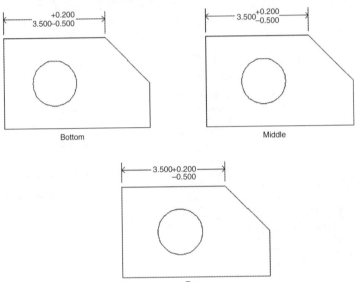

- If the **Deviation** method or **Limits** method is selected, then you have to choose whether to **Align decimal separators** or **Align operational symbols**, as shown below:

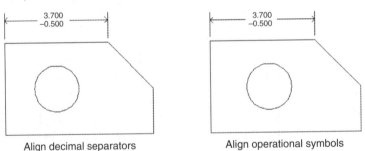

- Under Alternate units tolerances, if the **Alternate units** option is turned on, specify the **Precision** of the numbers. Consequently, choose **Zero suppression** for both the **Primary units** tolerance and the **Alternate units** tolerance.

16.10 CREATING A DIMENSION SUB-STYLE

- By default, the dimension style you create will affect all types of dimensions. If you want the dimension style to affect only a certain type of dimension and not the others you have to create a sub-style. Take the following steps:
  - Select an existing dimension style.
  - Use the **New** button to create a new style, and you will see the following dialog box:

  - Go to **Use for** and select the type of dimension (in the following example we selected Diameter) and the dialog box will change to:

  - Click the **Continue** button, and make the changes you want; these changes will affect diameter dimensions only.
  - The **Dimension Style** dialog box allows you to differentiate between the style and the sub-style and its features:

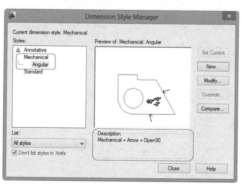

PRACTICE 16-1

Creating Dimension Styles

1. Start AutoCAD 2014.
2. Open **Practice 16-1.dwg**.
3. Create a new dimension style based on Standard, using the following information:
   a. Name: Part
   b. Extend beyond dim line = 0.3
   c. Offset from origin = 0.15
   d. Arrowhead = Right angle
   e. Arrow size = 0.25
   f. Center mark = Line
   g. Arc length symbol = Above dimension text
   h. Jog angle = 30
   i. Text placement vertical = Above
   j. Offset from dim line = 0.2
   k. Text alignment = ISO Standard
   l. Text placement = Over dimension line with leader
   m. Primary unit format = Fractional
   n. Primary unit precision = 0 ¼
   o. Fraction format = Diagonal
4. Click OK to end the creation process.
5. Select Part and create a sub-style for Radius, using the following:
   a. Arrowhead = Closed filled
   b. Arrow size = 0.15
6. Make Dimension the current layer.
7. Make Part the current dimension style, and add the dimensions to the shape to look like the following:

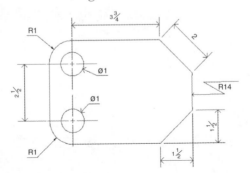

8. Save and close the file.

16.11 MORE DIMENSION FUNCTIONS

- In this section, we will discuss more dimension functions that will help you produce a better look for the final drawing. These functions are:
  - Dimension Break
  - Dimension Adjust Space
  - Dimension Jog Line
  - Dimension Center Mark
  - Dimension Oblique
  - Dimension Text Angle
  - Dimension Justify
  - Dimension Override

16.11.1 Dimension Break

- If two or more dimension blocks intersect in one or more points, this command will break one of the blocks at the intersection point. To issue this command, go to the **Annotate** tab, locate the **Dimensions** panel, and then select **Break** button:

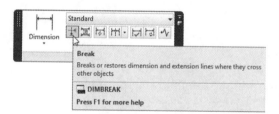

- You will see the following prompts:

```
Select dimension to add/remove break or [Multiple]:
Select object to break dimension or [Auto/Manual/ Remove]
<Auto>:
```

- The first prompt will ask you to select the dimension block that will be broken, and the second prompt will ask you to select the dimension block that will stay as is. These prompts will remove a break if it exists. At the second prompt, right-click and select the Remove option. These two prompts will be repeated until you press [Enter] to end the command.

- The final result will look like the following:

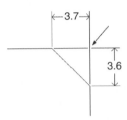

16.11.2 Dimension Adjust Space

- This command will help you adjust the spaces between dimension blocks to be either aligned or have equal spaces between them. To issue this command, go to the **Annotate** tab, locate the **Dimensions** panel, and then select the **Adjust Space** button:

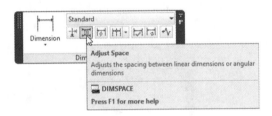

- You will see the following prompts:

```
Select base dimension:
Select dimensions to space:
Select dimensions to space:
Enter value or [Auto] <Auto>:
```

- The first step to select the base dimension block the other blocks will follow, then select all the other dimension blocks. When done, press [Enter], and AutoCAD will ask for the value; there are three options:
  - Value = 0 (zero), which means all the other blocks will be aligned with the base dimension block.
  - Value > 0, which will be the distance that will separate the base dimension block from the nearest block, and the others as well.
  - Value = Auto, which means AutoCAD will try to figure out the best arrangement for the selected blocks.

- See the following example:
  - We have the following situation:

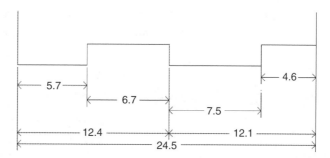

- As the first step, we started the Adjust Space command and selected the dimension block at the left reading 5.7 as our base dimension block, then we selected the adjacent 6.7, 7.5, and 4.6 and set the value to 0:

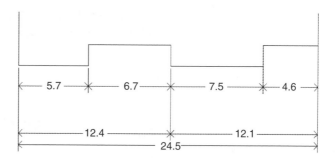

- We started the command again, and we selected the dimension block reading 5.7 as our base dimension block, then we selected the other three dimension blocks reading 12.4, 12.1, and 24.5 and setting the value to 1.0:

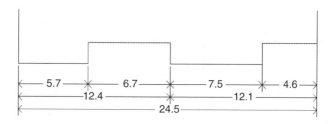

16.11.3 Dimension Jog Line

- This command will help you add/remove a jog line to an existing linear or aligned dimension. To issue this command, go to the **Annotate** tab, locate the **Dimensions** panel, and then select **Jog line** button:

- You will see the following prompts:

```
Select dimension to add jog or [Remove]:
Specify jog location (or press ENTER):
```

- The first prompt asks you to select the desired linear or aligned dimension block. You can select the position of the jog, or you can press [Enter] to let AutoCAD locate it automatically; you will see something like the following:

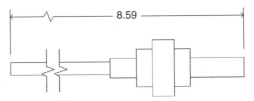

16.11.4 Dimension Center Mark

- This command will help you add a center mark to an existing circle/arc. To issue this command, go to the **Annotate** tab, locate the **Dimensions** panel, and then select **Center Mark** button:

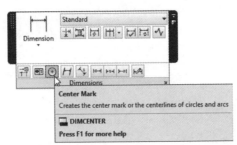

- You will see the following prompt:

```
Select arc or circle:
```

- Simply select the existing arc or circle, and a center will be added automatically.

16.11.5 Dimension Oblique

- This command will help you change the angle of the extension lines to any angle you want. To issue this command, go to the **Annotate** tab, locate the **Dimensions** panel, and then select **Oblique** button:

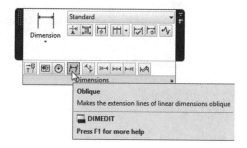

- You will see the following prompts:

```
Select objects:
Select objects:
Enter obliquing angle (press ENTER for none):
```

- Select the desired dimension block(s), then press [Enter]. Lastly, input the oblique angle, which can be positive or negative. See the following example:

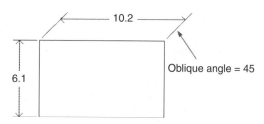

16.11.6 Dimension Text Angle

- This command will help you change the angle of the dimension text to any angle you want. To issue this command, go to the **Annotate** tab, locate the **Dimensions** panel, and then select **Text Angle** button:

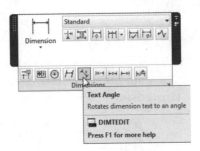

- You will see the following prompts:

```
Select dimension:
Specify angle for dimension text:
```

- Select the desired dimension block(s), and input the text angle, which can be positive or negative. See the following example:

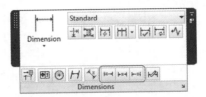

Text angle = 90

10.2

16.11.7 Dimension Justify

- This command will help you change the horizontal position of the dimension text. To issue this command, go to the **Annotate** tab, locate the **Dimensions** panel, and then one of the following buttons:

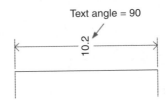

- Each of the functions will move the dimension text either to the left, center, to the right. See the following example (it moved it to the left):

16.11.8 Dimension Override

- This command will help you override a dimension system variable (you should memorize the system variable) or simply remove the override. To issue this command, go to the **Annotate** tab, locate the **Dimensions** panel, and then select the **Override** button:

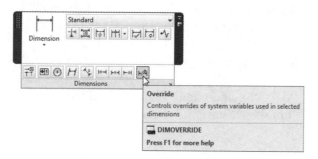

- You will see the following prompts:

```
Enter dimension variable name to override or [Clear
overrides]:
```

- This command is very useful if you want to remove (clear) all the overriding steps you take on a dimension block using the right-click menu. To answer to the above prompt, either type the name of the dimension system variable, or type C to clear the override. See the following illustration:

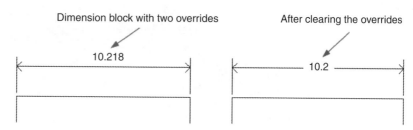

Dimension block with two overrides   After clearing the overrides

10.218   10.2

PRACTICE 16-2

More Dimension Functions

1. Start AutoCAD 2014.
2. Open **Practice 16-2.dwg**.
3. Using Adjust Space adjust the space between the six continuous dimensions to be all at the same alignment with 2¾ dimension.
4. Using Adjust Space adjust the space between the six continuous dimensions and the total single dimension to 1.5.
5. Make sure that Dimension layer is the current layer; if not, make it current.
6. Add center marks to the two circles.
7. Using Dimension Break, break the horizontal 1½ using the vertical 1½.
8. Rotate the text of the vertical 1½ to be 45.
9. What does the upper horizontal dimension read? _____
10. Using Dimension Override, clear the override. What does it read now? _____.
11. Make the same dimension left justified.
12. You should have the following:

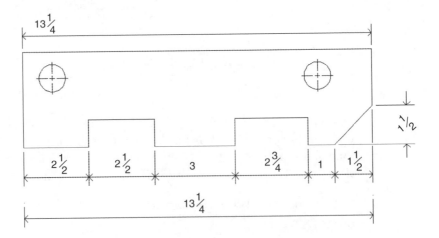

13. Save and close the file.

16.12 HOW TO CREATE A MULTILEADER STYLE

- A multileader is a replacement for the normal leaders that used to exist in AutoCAD. Leaders used to follow the current dimension style, and they were always single ones. A multileader has its own style, and a single leader can point to a different location in the drawing.
- A multileader style allows you to set the characteristics of a multileader block. To start this command, go to the **Annotate** tab, locate the **Leaders** panel, and then select the **Multileader Style** button:

- You will see the following dialog box:

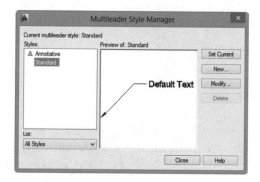

- As you can see there are two predefined styles: one called Standard (the default) and the other called Annotative. Click the **New** button to create a new multileader style, and you will see the following dialog box:

■   Input the name of the new style, and then click the **Continue** button.
    There are three tabs, and each one will control part of the multileader
    block:
    • Leader Format
    • Leader Structure
    • Content

16.12.1 Leader Format Tab

■   This is the Leader Format tab:

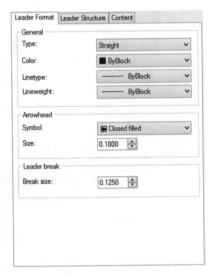

■   Here, you can change all or any of the following:
    • Edit the **Type** of the leader, and choose one of the three choices:
      Straight, Spline, or None. Here is an example of both the Straight and
      Spline options:

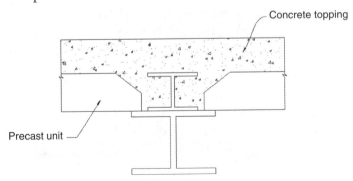

- Edit the **Color**, **Linetype**, and **Lineweight**.
- Select the **Arrowhead** shape and its size.
- Set the distance of the dimension break from any two blocks that intersect.

16.12.2 Leader Structure Tab

- This is the Leader Structure tab:

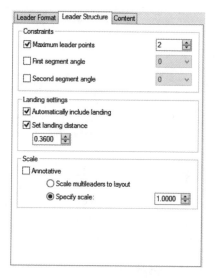

- You can change all or any of the following:
  - Specify if you want to change the **Maximum leader points**, and then input the desired value. By default, this value is 2; that is, the first point points to the geometry and the second point is the end of the multileader:

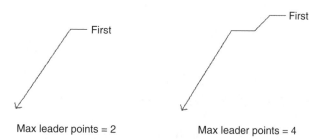

Max leader points = 2   Max leader points = 4

  - Specify whether you want to change **First segment angle** and **Second segment angle**. If yes, specify the angle values.

- Specify whether you want AutoCAD to **Automatically include landing**. If yes, specify the **landing length**.
- Specify whether the multileader will be **Annotative** (discussed in Chapter 9).

16.12.3 Content Tab

- This is the Content tab:

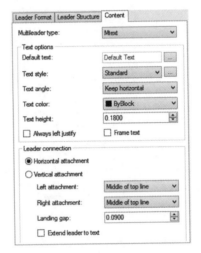

- In AutoCAD, there are two multileader types:
  - Mtext
  - Block (either pre-defined or user-defined)
- The following illustration shows the two types:

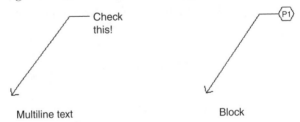

Multiline text   Block

- If you select the **Mtext** option, you will be able to change all or any of the following:
  - If there is **Default text**.
  - **Text style**, **Text angle**, **Text color**, and **Text height** (if Text style's height = 0).

- Whether the text is **Always left Justify** and with **Frame**.
- Whether the leader connection is horizontal or vertical. If vertical, then edit the position of the text relative to the landing for both left and right leader lines, and then control the gap distance between the end of the landing and the text:

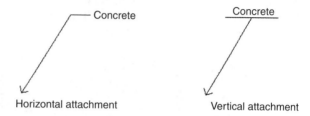

Horizontal attachment   Vertical attachment

- If you select the **Block** option, you can change all or any of the following:

- Specify the **Source block**:

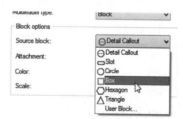

- Specify the **Attachment** position, **Color** of the attachment, and finally, the **Scale** of the attachment.

16.13 INSERTING A MULTILEADER DIMENSION

- This group of commands allows you to add a single multileader, add a leader to an existing multileader, remove a leader from an existing multileader, and align and group an existing multileader. You will always start with the Multileader command, which will insert a single leader. To start this command, go to the **Annotate** tab, locate the **Leaders** panel, and then select the **Multileader** button:

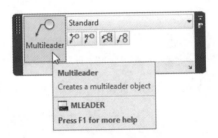

- You will see the following prompts:

```
Specify leader arrowhead location or [leader Landing
first/Content first/Options] <Options>:
Specify leader landing location:
```

- First, specify the leader arrowhead location, then specify the leader landing location, and then type the text you want to appear beside the leader. To add a leader to an existing multileader, go to the **Annotate** tab, locate the **Leaders** panel, and then select the **Add Leader** button to add more leaders:

- You will see the following prompts:

```
Select a multileader:
1 found
```

```
Specify leader arrowhead location:
Specify leader arrowhead location:
```

- To remove a leader from an existing multileader, go to the **Annotate** tab, locate the **Leaders** panel, and then select the **Remove Leader** button:

- You will see the following prompts:

```
Select a multileader:
1 found
Specify leaders to remove:
Specify leaders to remove:
```

- To align a group of multileaders, go to the **Annotate** tab, locate the **Leaders** panel, then select the **Align** button:

- You will see the following prompts:

```
Select multileaders: 1 found
Select multileaders: 1 found, 2 total
Select multileaders:
Current mode: Use current spacing
Select multileader to align to or [Options]:
Specify direction:
```

- To collect a group of similar multileaders to be a single leader, go to the **Annotate** tab, locate the **Leaders** panel, and then select the **Collect** button. This command works only with leaders containing blocks:

- You will see the following prompts:

```
Select multileaders:
Select multileaders:
Specify collected multileader location or
[Vertical/Horizontal/Wrap] <Horizontal>:
```

PRACTICE 16-3

Creating Multileader Styles and Inserting a Multileader

1. Start AutoCAD 2014.
2. Open **Practice 16-3.dwg**.
3. Create a new multileader style based on <u>Standard, using the following</u> <u>information</u>:
   a. Name = Texture and Painting
   b. Arrowhead symbol = Dot small
   c. Arrowhead size = 0.35
   d. First segment angle = 0
   e. Automatically include landing = Off
   f. Multileader type = Block
   g. Source block = Circle
4. Create a new multileader style based on <u>Standard, using the following</u> <u>information</u>:
   a. Name = Material
   b. Leader format = Spline
   c. Arrowhead symbol = Right angle
   d. Arrowhead size = 0.25
5. Make layer Dimension current.

6.   Using both styles insert the following multileaders:

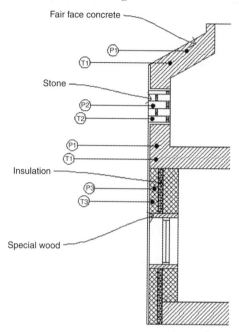

7.   Using Add leader, Align, and Collect try to get the following:

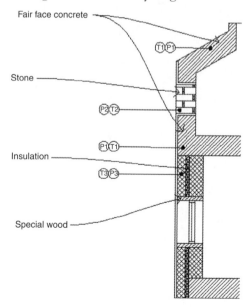

8.   Save and close the file.

NOTES:

CHAPTER REVIEW

1. Symmetrical and Deviation are two types of _____.
2. There are two types of multileader blocks.
   a. True
   b. False
3. When creating a new dimension style, which of the following is *not* correct?
   a. You can create dimension style affecting all types of dimensions.
   b. You have to select the existing dimension style to begin with.
   c. You can't create a dimension style affecting only one type of dimension.
   d. You can create a sub-style.
4. You can show _____ units and _____ units in a dimension block.
5. When using a multileader style there should always be a landing in your block.
   a. True
   b. False
6. Collect and Align are _____ commands.

CHAPTER REVIEW ANSWERS

1. Tolerance
3. c
5. b

17

PLOT STYLE, ANNOTATIVE, AND DWF FILES

In This Chapter

◇ How to create and use the two types of plot styles
◇ Using the Annotative feature in AutoCAD
◇ Creating and viewing DWF files

17.1 PLOT STYLE TABLES – FIRST LOOK

■ Plot styles are used to convert the colors used in the drawing to printed colors. The default setting is to keep the same color as the printer. Before AutoCAD 2000, there was only one type of conversion method, but more recent versions of AutoCAD have a new feature called Plot Style Tables. There are two types of Plot Style Tables:
  • Color-dependent Plot Style Table
  • Named Plot Style Table

17.2 COLOR-DEPENDENT PLOT STYLE TABLES

■ This plot style table is a simulation for the only method that existed before AutoCAD 2000. The essence of this method is simple: for each color used in your drawing you will specify the color to be used at the printer. AutoCAD allows you to set the lineweight, linetype, etc., for each color. The problem with this method is that you only have 255 colors to choose from.

- Each time you create a Color-dependent Plot Style Table, AutoCAD will create a file with the extension *.ctb*. You can create Plot Style Tables from outside AutoCAD using the Control Panel in Windows, or from inside AutoCAD using the menu bar.
- From outside AutoCAD, start the Control Panel in Windows, double-click the **Autodesk Plot Style Manager** icon, or show the menu bar, and then select **Tools/Wizards/Add Plot Style Table**. You will see the following dialog box:

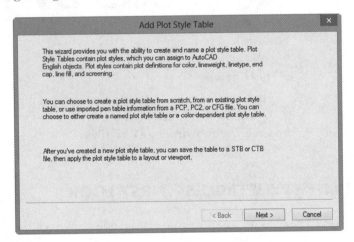

- The first screen is an introduction to the whole concept of plot styles; read it to understand the next steps. When done, click the **Next** button, and you will see the following dialog box:

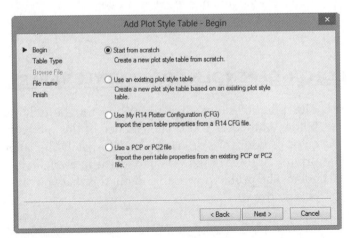

- AutoCAD will give you four choices:
  - Start from scratch
  - Use an existing plot style
  - Import AutoCAD R14 CFG file and create a plot style from it
  - Import PCP or PC2 file, and create a plot style from it
- Select **Start from scratch**, click the **Next** button, and you will see the following dialog box:

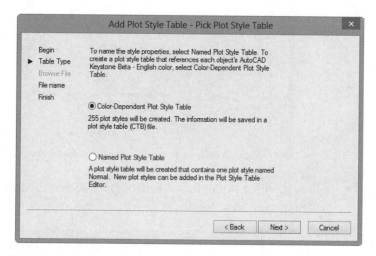

- Select **Color-Dependent Plot Style Table**, click the **Next** button, and you will see the following dialog box:

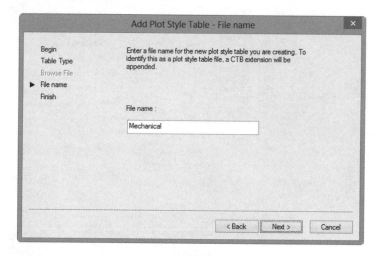

■ Input the name of the new plot style, click the **Next** button, and you will see the following dialog box:

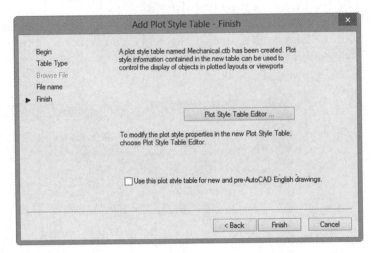

■ Select the **Plot Style Table Editor** button, and the following dialog box will be displayed:

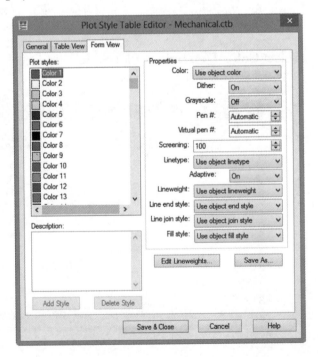

- From the left, select the color you used in your drawing file, and then on the right, change all or any of the following settings:
  - Change the **Color** to be used in the plotter.
  - Switch **Dither** on/off. This option will be dimmed if your printer or plotter doesn't support dithering. Dither is a method used to give the impression of more colors than the 255 colors used by AutoCAD. It is better to leave this option off, but it should be on if you want **Screening** to work.
  - Turn **Grayscale** on/off. This option is good for laser printers.
  - Change **Pen #**. This option is valid for the old types of plotters – pen plotters – that are no longer used.
  - Change **Virtual pen #**. Used for non-pen plotters to simulate pen plotters by assigning a virtual pen for each color; leave it on **Automatic**.
  - Change **Screening**. This option will reduce the intensity of the shading and fill the hatches, hence reducing the amount of ink used. This option depends on **Dither**.
  - Change **Linetype**. Set a different linetype for the color or leave it as the object's linetype.
  - Change **Adaptive**. This option will change the linetype scale of all objects using the color to start a segment and end a segment. Turn this option off if the linetype scale is important for your drawing.
  - Change **Lineweight**. This option will change the lineweight for the color selected.
  - Change **Line end style**. This option allows you to select the end style for lines; choose one of the following: Butt, Square, Round, and Diamond.
  - Change **Line join style**. To select the line join shape, choose from Miter, Bevel, Round, and Diamond
  - Change **Fill style**. This option will set the fill style for the filled area in the drawing (good for trial printing).
- Click **Save & Close**. Then click **Finish**.
- Your last step should be linking your plot style with a layout. Take the following steps:
  - Select the desired layout, then start the **Page Setup Manager**.
  - Select the name of the current page setup and click **Modify**.

- At the upper-right of the dialog box, under **Plot style table (pen assignments)**, select the desired plot style table:

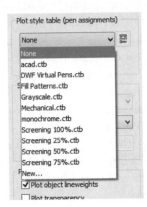

- Click the **Display plot styles** checkbox on.
- For each layout, you can assign one *ctb* file. In order to see the linetype and lineweight of the objects, you have to click the **Show/Hide Lineweight** button on the status bar on, as shown below:

17.3 NAMED PLOT STYLE TABLES

- This method does not depend on colors. The created plot style tables will be linked later on with the layers, so you may have two layers with the same color, yet they will print with different colors, linetypes, and lineweights.
- The Named Plot Style Table has the file extension **.stb.* The procedure for creating a Named Plot Style is identical to creating a Color-dependent Plot Style, except the last step, which is configuring the **Plot Style Table Editor**. You can create it from outside AutoCAD using the Control Panel and double-clicking the **Autodesk Plot Style Manager** icon, or you can show the menu bar and then select **Tools/Wizards/Add Plot Style Table**. You will see the same screen you saw while creating a Color-dependent Plot Style Table.

- Follow the same steps until you reach the **Plot Style Table Editor** button. Click it and you will see the following dialog box. Click the **Add Style** button:

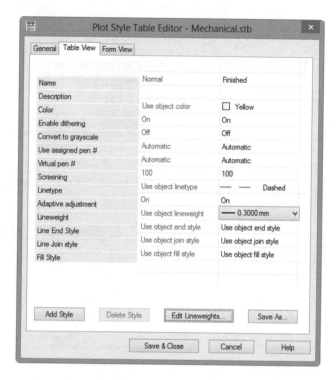

- As you can see you can change all or any of the following:
  - Input the **Name** of the style and a brief **Description**.
  - Change the **Color** to be used in the plotter.
  - We explained the rest of features in the previous section on Color-dependent Plot Style Tables.
- You can add as many styles as you wish in the same Named Plot Style. Click **Save & Close**. Then click **Finish**.
- Linking a Named Plot Style Table with any drawing is a bit more complicated than linking a Color-dependent Plot Style Table:
- The first step is a precautionary step. You may want to print a drawing, but then discover you can only use .ctb files. To solve this problem you have to convert one of the .ctb files to a .stb file. At the Command window type **convertctb**, and a dialog box listing all the .ctb files will

appear. Select one of them, keeping the same name, or give it a new name, and then click **OK**. You will see the following dialog box:

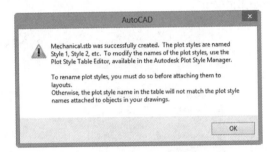

- Convert the drawing from a Color-dependent Plot Style to a Named Plot Style. At the Command window type **convertpstyles**, and you will see the following warning message:

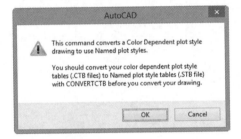

- Click **OK**, and you will see the following dialog box:

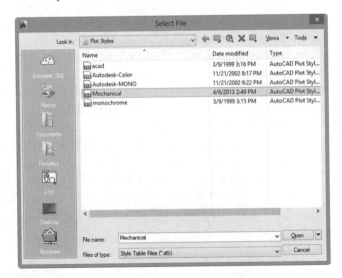

- Select the newly created Named Plot Style Table, click **Open**, and you will see the following prompt:

```
Drawing converted from Color Dependent mode to
Named plot style mode.
```

- The second step is to select the desired layout. Start the **Page Setup Manager**, and at the upper-right of the dialog box, under **Plot style table (pen assignments)**, select the name of the newly created Named Plot Style Table. Click the **Display plot styles** checkbox on, and end the Page Setup Manager command:

- Select the **Layer Properties Manager**, and for the desired layer(s) click the name of the Plot Style in the **VP Plot Style** column:

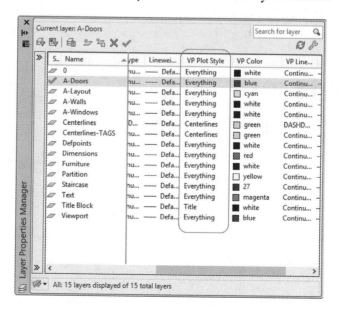

- You will see the following dialog box:

- Select the desired plot style. When done, click **OK**.
- Repeat the same steps for the other layers.
- You may need to type **regenall** at the Command window (which means regenerate all viewports) in order to see the effects of your changes.

 ■ If you create a new drawing using *acad.dwt* your drawing will only accept a Color-dependent Plot Style Table. Use *acad-Named Plot Styles.dwt* to create a new drawing that will only accept a Named Plot Style Table.

PRACTICE 17-1

 Color-dependent Plot Style Tables

1. Start AutoCAD 2014.
2. Open **Practice 17-1.dwg**.
3. Create a new Color-dependent Plot Style Table and name it Architectural using the following settings:

Color	1	2	3	4	5	6	7	27
Plot color	Black	Black	Use object color	Black	Black	Black	Black	Black
Lineweight	0.3	0.3	0.3	0.3	0.7	0.3	0.3	0.3

4.   Switch to the D-Size Architectural Plan layout.
5.   Link this layout to the Architectural Plot Style.
6.   Check that the Show/Hide Lineweight button on the status bar is turned on.
7.   Compare the doors with the other objects, and you will find they are thicker, because color (5) was used for both the title block and doors with a 0.7 lineweight.
8.   Save and close the file.

PRACTICE 17-2

Named Plot Style Tables

1.   Start AutoCAD 2014.
2.   Open **Practice 17-2.dwg**.
3.   Create a new Named Plot Style Table and name it Architectural using the following settings:

Style name	Color	Lineweight
Everything	Black	0.3
Centerlines	Green	0.5
Title	Black	0.7

4.   Link the newly created table with the D-Size Architectural Plan layout.
5.   Using the Layer Properties Manager make the following changes:
     a.   Layer = Centerlines and Centerlines-TAGS linked to Centerlines
     b.   Title Block linked to Title
     c.   The rest of the layers linked to Everything
6.   Zoom in to compare the Centerlines lineweight to the other objects.
7.   Save and close.

17.4 WHAT IS THE ANNOTATIVE FEATURE?

- Since we will always print from layout, we need to use viewports, and since we will use viewports we have to set the viewport scale for each viewport. The viewport scale will affect all objects in Model space. So, if you hatch, type text, insert dimensions, insert a multileader, or insert a block contains text in Model space, all of these objects will be scaled. If they are scaled down, then the text and dimension will be unreadable, and the hatch will look like solid hatching.
- What we need is a feature which will scale everything except the annotation objects (hatches, text, and dimensions), and this feature is called the **Annotative** feature.
- The Annotative feature can be found in different places:
  - You will find it in Text style, under Size:

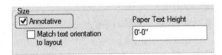

  - You will find it in Dimension style in the Fit tab:

  - You will find it in Multileader style in the Leader Structure tab:

  - You will find it in the Hatch context tab in the Options panel:

- You will find it in Block creation command, under Behavior:

■ How will you know that you are dealing with a style that supports the Annotative feature? You will see a special symbol by its name. See the illustration below:

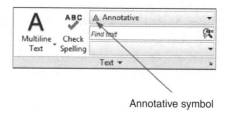

Annotative symbol

■ How will you know that an object was inserted using a style or a command that supports the Annotative feature? Simply hover over it, and if you see something like the following, you know it was inserted with the Annotative feature on:

This is my first Annotative text

■ In order to work with annotative annotation objects, take the following simple steps:
  - Create your drawing in Model space, without any annotation objects (hatches, text, dimensions, multileaders, and blocks with text).
  - Select the desired layout, add viewports, and then scale them. This scale will be for both the annotation objects and the viewport.
  - Double-click inside the viewport to make it active.
  - Insert all the annotation objects you need.

- Once you insert annotation objects in the viewport, you can see them in this viewport and any other viewport with the same scale value.
- If you change the scale of the viewport, you will lose the annotation object.
- To show the annotation object in more than one viewport with different scale values, right click on the **Annotation Visibility** button (at the right portion of the status bar), and you will see the following menu:

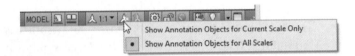

- Select either to **Show Annotation Objects for Current Scale Only** or **Show Annotation Objects for All Scales**.
- Control how to add a scale values to the viewport: **Automatic** or **Manual**. Right-click on the **Add Scale** button (at the right portion of the status bar), and you will see the following menu:

- If you use a zooming command inside the viewport you will ruin the current viewport scale. To get it back click the **Synchronize** button (at the right portion of the status bar), and you will see the following message (use the lock in the status bar to lock the viewport scale to avoid this problem):

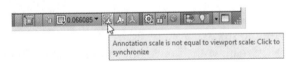

- Clicking this message will restore the viewport scale.

- To make an annotation object appear in viewports with different scales, select it, right-click, and select the **Annotative Object Scale** option. You will see the following sub-menu:

- Choose the **Add/Delete Scales** option, and the following dialog box will appear:

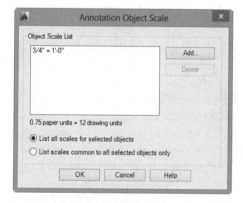

- Select the **Add** button, and the following dialog box will appear:

- Select the desired scale value, and click **OK** twice.

PRACTICE 17-3

Annotative Feature

1. Start AutoCAD 2014.
2. Open **Practice 17-3.dwg**.
3. Make layer Dimensions current.
4. Switch to the Details layout.
5. There are three viewports. The first one at the left is scaled to 1:20, and the other two at the right are scaled to 1:10.
6. Go to the Annotate tab, locate the Dimensions panel, and check the available dimension styles. You will see only one with a distinctive symbol at its left, which is mechanical. This dimension style is annotative.
7. Double-click inside the big viewport.
8. Start the Radius command, select the big magenta circle, and then add the dimension block. Make sure it didn't appear in the upper-right viewport.
9. Add another Radius block to the small magenta circle.
10. Add two linear dimensions for the total width and the total height.
11. While still inside the same viewport, make layer Text current.
12. Make the text style Annotative current.
13. Using Multiline text add the word Bearing beneath the shape.
14. Make layer Dimensions current again.
15. Click inside the upper-right viewport to make it current and add a Radius dimension to one of the small circles.
16. Make layer Hatch current.
17. Click inside the lower-right viewport.
18. Start the Hatch command, make sure that Annotative is on, and set the scale = 10. Hatch the area between the two dashed lines and then end the Hatch command.
19. Using Add/Delete scales add scale 1:20 so the two hatches will appear in the big viewport.

20. You should have the following:

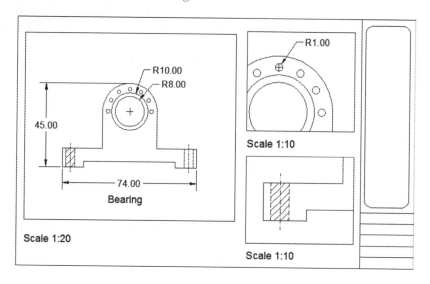

21. Save and close the file.

17.5 DESIGN WEB FORMAT (DWF) FILES

- Sharing files and data is important for almost everyone in today's world. But if you send your .dwg files your design could be stolen. File size is also an issue; .dwg files are large (the file size may be more than 50–100 MB) and are very difficult to send via e-mail. Additionally, other users need AutoCAD to open these files. To solve these problems, AutoCAD allows us to plot to the **D**esign **W**eb **F**ormat (DWF) file type. A DWF file can't be modified, so you don't have to worry about your design being stolen, and the file size is relatively small compared to the DWG format. To view a DWF file you need a free program called Autodesk Design Review, which comes on the same DVD as AutoCAD, or it can be installed from the Autodesk Web site. This software is for viewing and printing, but you can also use it to measure and red-line DWF files as well.
- Another version of the DWF file type, called DWFx, can be viewed with both Windows Vista and Windows 7 using the Internet Explorer browser.

17.6 EXPORTING DWF, DWFX, AND PDF FILES

- This command allows you to export your current DWG file to DWF, DWFx, and PDF formats. To start this group of commands, go to the **Output** tab, locate the **Export to DWF/PDF** panel, and then select the **Export** button:

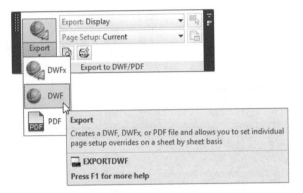

- You will see the following dialog box:

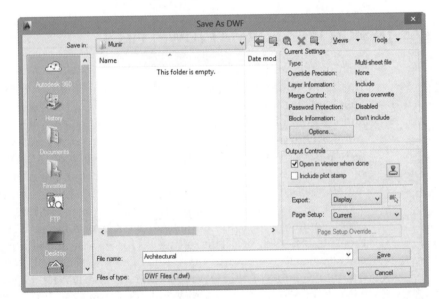

- At the top-right of the dialog box, under **Current Settings**, AutoCAD lists the current settings, as shown below:

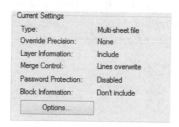

- To edit these settings click the **Options** button, and the following dialog box will appear:

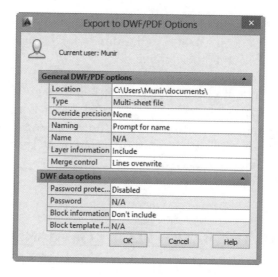

- Change all or any of the following settings:
  - Indicate the location on your computer where you will save the DWF file.
  - File **Type** by default is always Multiple-sheet.
  - Indicate the **Override** precision; you can choose from Manufacturing, Architectural, etc. This setting allows you to choose the dpi (dot per inch) precision, which allows DWF users to measure distances accurately.
  - Input the **Name** of the DWF file, or let AutoCAD prompt you after executing the command.

- Indicate whether to include **Layer** information. This allows you to turn off layers in the Autodesk Design Review software.
- Indicate whether to include a **Password**.
- To finish, click **OK**.
■ Under **Output Controls** you will see the following:

■ Set all or any of the following:
- To open a DWF using Autodesk Design Review automatically after executing the command.
- To include a plot stamp.
- Indicate what to export. If you execute the command from Model space, then select Display, Extents, or Window. On the other hand, if you are exporting while you are at one of the layouts, then the available selections will be Current layout or All layouts.
- Select the Page Setup.
- To finish, click the **Save** button.

17.7 USING THE BATCH PLOT COMMAND

■ This command allows you to produce a DWF file containing multiple layouts from the current drawing and from other drawings. To issue this command, go to the **Output** tab, locate the **Plot** panel, and then select the **Batch Plot** button:

- You will see the following dialog box:

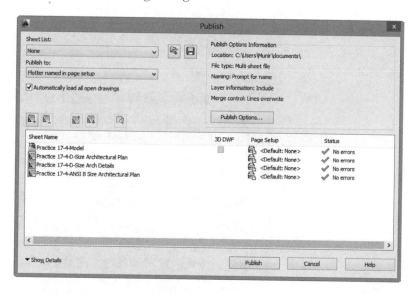

- You will see a table contains Model space and layouts. Set all or any of the following:
  - Indicate **Publish to** either the printer/plotter defined in the layout or DWF, DWFx, PDF.
  - Indicate whether to load all open drawings automatically.
  - Using the five buttons above the table, you can add a sheet, remove a sheet, and move any sheet up or down to specify its order relative to the document sheets. Finally, you can preview the sheet:

  - You can rename the sheets by clicking on any sheet name, and the name will become editable.

- Click **Publish Options**, and the following dialog box will appear, which is identical to what we discussed in the Exporting section:

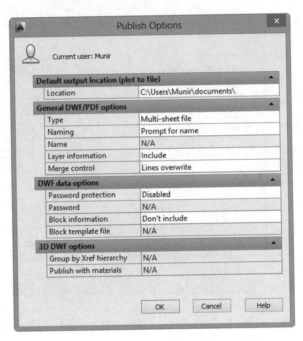

- When done click **OK**.
- Specify the number of copies.
- Select whether to include a plot stamp.
- Indicate whether you want to publish a DWF in the background.
- Indicate whether to open the DWF using Autodesk Design Review.
- Set the precision for the DWF file (dpi precision) for measurements in Autodesk Design Review.
- To finish, click the **Publish** button. A final message will come up as a bubble similar to the following:

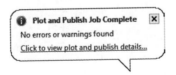

17.8 VIEWING DWF AND DWFX FILES

- AutoCAD comes with Autodesk Design Review. You will see a shortcut on your Desktop after you install AutoCAD. Once you locate it, start it, and open a DWF or DWFx file (you can only open one file at a time), you will see the following:

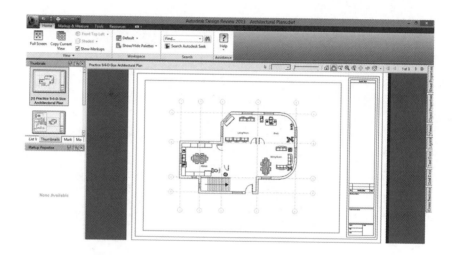

PRACTICE 17-4

Creating and Viewing DWF Files

1. Start AutoCAD 2014.
2. Open **Practice 17-4.dwg**.
3. Produce a multisheet DWF file including all layouts except the Model space sheet. Make sure to include layers in the DWF file. Save the file as *Architectural Plans.dwf*. Open it using Autodesk Design Review.
4. If you have Windows Vista or Windows 7, produce a DWFx file, and open it using Internet Explorer.
5. Save and close the file.

NOTES:

CHAPTER REVIEW

1. Color-dependent Plot Styles are similar to those used before AutoCAD 2000.
   a. True
   b. False
2. The best practice is to insert dimensions, text, and hatches in _____, using the _____ feature.
3. A _____ is better than a Color-dependent Plot Style Table.
4. There is no difference between DWF and DWFx files.
   a. True
   b. False
5. In order to insert an annotative object, you have to be inside the viewport.
   a. True
   b. False

CHAPTER REVIEW ANSWERS

1. a
3. Named Plot Style Table
5. a

18

HOW TO CREATE A TEMPLATE FILE AND INTERFACE CUSTOMIZATION

In This Chapter

◇ How to create a template file
◇ How to use the CUI command to customize AutoCAD

18.1 WHAT IS A TEMPLATE FILE AND HOW DO YOU CREATE ONE?

- Any company using AutoCAD should standardize their work and shorten their production time using templates whenever possible. Standardization includes using the same layer naming, colors, linetype, and lineweight as well as standard text and tables, dimensions and leaders, and standard layouts as well as the same shapes for blocks. Templates store this information.
- To create a good template, you should include the following:
  - Drawing units
  - Drawing limits
  - Grid and Snap settings
  - Layers
  - Linetypes
  - Text Styles
  - Table Styles
  - Dimension Styles
  - Multileader Styles
  - Layouts (including Border blocks, and Viewports)
  - Page Setups
  - Plot Style tables

- ■ Take the following steps to create a template file:
  - Create a new file using template file **acad.dwt** or **acadiso.dwt**.
  - Prepare the above settings.
  - Create inside this file all the above settings.
  - From the Application menu choose **Save As/AutoCAD Drawing Template**:

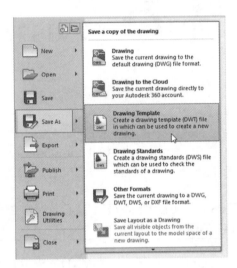

- ■ You will see the following dialog box:

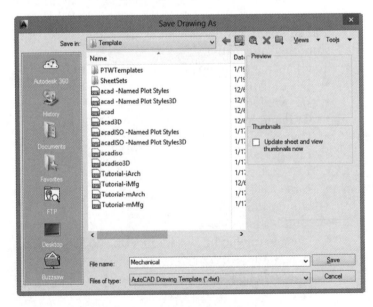

- Input the name of the template file. By default AutoCAD will direct you to the same folder it uses to save its default template files. You can save your template here, or you can create your own folder, which we highly recommend. But to use this method you have to specify the new folder for AutoCAD. You can do this using the Options dialog box. See the following illustration:

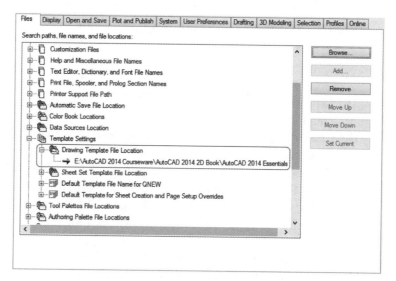

18.2 EDITING A TEMPLATE FILE

- To edit an existing template do the following:
  - Use the normal Open file command.
  - You will see the following dialog box. Using **Files of type** choose **Drawing Template (*.dwt)**:

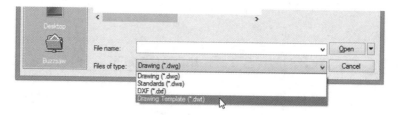

- It will take you to the default Template folder. Select and open the desired template, and then make your edits.
- Save it with the same name, or a new name.

PRACTICE 18-1

Creating and Editing Template File

1. Start AutoCAD 2014.
2. Open **Practice 18-1.dwg**.
3. Go to the Model and erase all objects.
4. Delete the D-Size Arch Details layout.
5. Go to the ANSI B Size Architectural Plan layout and delete the circular viewport.
6. Rename ANSI B Size Architectural Plan to be ANSI B Size.
7. Go to the D-Size Architectural Plan layout and delete the viewport.
8. Rename D-Size Architectural Plan to ANSI D Size.
9. Go to the Annotate tab and check the existing dimension style and text style.
10. Go to the Home tab and check the existing layers.
11. Go to the Application menu, start the Units command, and set the precision for length to be 0'-0 ½". Click OK to end the command.
12. Save the file as a template file under the name My Company.dwt (it will be saved in the Template folder where AutoCAD saves all templates). You can save the template file in different folder, but it should be specified in the Options dialog box in the Files tab.
13. Close the file.
14. Start a new file using the My Company.dwt template file, and you will find that all of your settings are there in the new file.
15. Close the new file without saving.

18.3 CUSTOMIZING THE INTERFACE – INTRODUCTION

- You can change the ribbons and panels of the interface to fit your own needs. You can also add a new workspace, new tabs, new panels, and even new commands using a single command: CUI (Customization User interface).

- To issue this command, go to the **Manage** Panel, locate the **Customization** panel, and then select the **User Interface**:

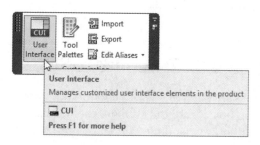

- You will see the following dialog box:

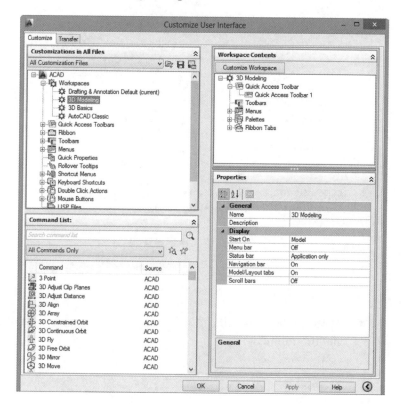

- This dialog box is divided into four parts:
  - Customization part: Specifies what you want to customize: Workspace, Tab, Panel, Toolbar, etc. This part allows you to create new, delete existing, rename, etc.

- Command list contains all AutoCAD commands.
- Depending on what you choose from the left, the right will change to Content and Properties.

18.4 HOW TO CREATE A NEW PANEL

- Panels are parts of tabs. To create a new panel, at the upper left of the CUI dialog box, locate the Ribbon, then expand it, and you will find that, with or without expanding, you can right-click Panels just like the following. Select the **New Panel** option:

- Type the name of the new panel, press the [Enter], and you will see something like the following:

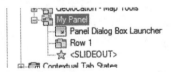

- At the right, you will see the Panel Preview with a small rectangle just like the following:

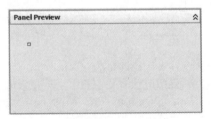

- You will see the Properties section, which will look like the following:

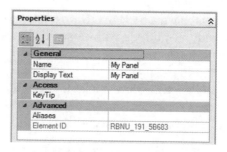

- When you create a new panel three things will automatically be added. You will see the following:

  - Panel Dialog Box Launcher
  - Row 1 (to add commands to this row or any other row, use the Command List section and drag-and-drop the desired command)
  - SLIDEOUT
- To better understand these three things, see the following illustration:

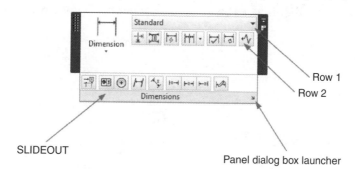

- To create a Panel decide on the following:
  - How many rows you will have
  - How many sub-panels you will have
  - The commands to be included and where (in rows or sub-panels)
  - The command presented as the large icon at the left

- If it will have a Panel Dialog Box Launcher, and what it is (normally it is a style or palette)
- If it will have a SLIDEOUT or not, and the commands in it
- By default all buttons will be small. To make a button large, take the following steps:
    - Drag it to the panel.
    - From Panel Preview, click the desired button.
    - At the Properties, check the Button Style and select Large (either with Text horizontal, vertical, or without text) just like the following:

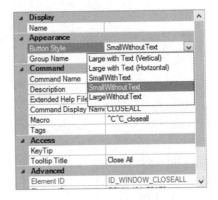

- To add a new row in a panel, do the following:
    - Select the desired panel.
    - Right-click the name and select the **New Row** option:

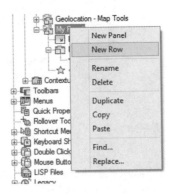

- Sub-panels, as the name suggests, separate the panel into smaller parts. Each sub-panel will have row1, row2, etc. To add a new sub-panel take the following steps:

- Select the desired row.
- Right-click and select New Sub-Panel:

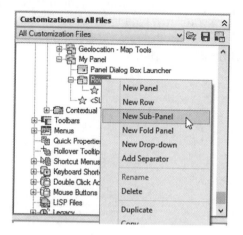

■ To add a vertical separator between buttons, take the following steps:
  - Select the desired row.
  - Right-click and select the **Add Separator** option:

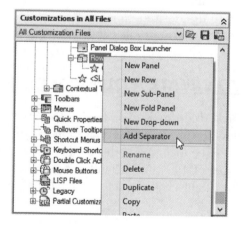

NOTE ■ Any row located below the slideout will be shown in it.

18.5 HOW TO CREATE A NEW TAB

- Tabs consist of panels. You can include panels included with AutoCAD, or create your own panels, or mix and match.
- To create a new tab, start the CUI command, and at the upper left expand the Ribbon, then right-click Tabs, and select the **New Tab** option:

- Type the name of the new tab.
- You should fill the tab with panels. Use one of the following methods:
  - Drag-and-drop: this method is not practical, as there is a huge list of panels and tabs.
  - Go to the desired panel, right-click and select Copy, then go to the desired tab, right-click and select Paste.

18.6 HOW TO CREATE A QUICK ACCESS TOOLBAR

- A Quick Access toolbar is a toolbar that appears at the top left of AutoCAD window to the right of Application menu. You can specify which commands to be included in this toolbar. To add a new Quick Access Toolbar, start the CUI command, and at the upper left, locate the Quick Access Toolbar, right-click and select the **New Quick Access Toolbar** option:

- Type the name of the new Quick Access Toolbar. By default six commands will be included: New, Open, Save, Undo, Redo, Plot. You can drag-and-drop any new command from the Commands List.

18.7 HOW TO CREATE A NEW WORKSPACE

- A workspace is a set of tabs (hence panels) that will appear together along with palettes, menus, toolbars, and Quick Access toolbars. To create a new workspace, start the CUI command, and at the upper left, locate Workspaces, right-click, and select the **New Workspace** option:

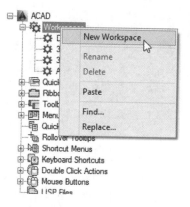

- Type the name of the new workspace. Click the newly created workspace, and look to the upper right. You will see something like the following:

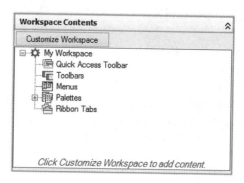

- As you can see, there are five different things to be included inside a workspace: Quick Access Toolbar, Toolbars, Menus, Palettes, and Ribbon Tabs. In order to add/remove the contents of the workspace, click the button at the top of the window, titled "Customize Workspace," and accordingly text will change to blue and at the upper left, you will see everything with a checkbox to its left, something like the following:

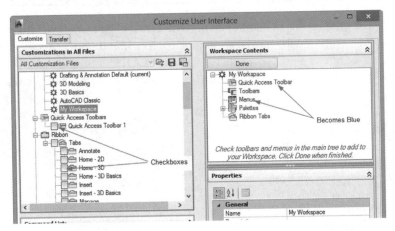

- So if you want to add anything to your workspace, simply click on the checkbox to copy it. When you are through with your adding, click "Done" to finish the process of customizing your workspace.

PRACTICE 18-2

Customizing Interface

1. Start AutoCAD 2014.
2. Start a new file.
3. Start the CUI command.
4. Create a new workspace and name it My Workspace.
5. Go to the Ribbon, expand it, and locate the Tabs.
6. Create a new tab and name it My Tab, then collapse the list.

7. Locate Panels, then create a new panel and name it My Panel. Take the required steps to make it look like the following:

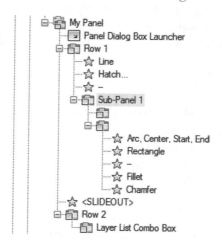

8. The Panel Dialog Box Launcher = Tool Palette command.
9. Copy My Panel to My Tab.
10. Add the following premade panels to My Tab:
    a. Annotate – Dimensions
    b. View – Viewports
    c. Home – Properties
11. Select My Workspace, then right-click the Customize Workspace button.
12. At the left click My Tab to be the only tab included in this workspace (you can include other premade tabs if you like).
13. Collapse Tabs.
14. Click Quick Access Toolbars and then Quick Access Toolbar 1.
15. Click OK to end the CUI command.
16. Go to the top of the screen and select My Workspace to make it current.
17. Test the panel you created.
18. Restore the Drafting and Annotation workspace.
19. Close the file without saving.

18.8 HOW TO CREATE YOUR OWN COMMAND

- In this section, we will learn how to create a new command by creating a macro, using AutoCAD commands with certain values, then assigning this macro to a button, and put it in a panel (hence in a tab).
- We will use certain characters in the macro:
    - ^C^C, which is equal to pressing [Esc] twice. Previously AutoCAD used [Ctrl] + C to cancel commands, then later on (AutoCAD R12) changed it to pressing [Esc] instead. We have to start each macro with ^C^C in order to make sure the macro will cancel all running commands before it starts.
    - Use ";" to simulate [Enter].
    - Use "\" to wait for the user input.
    - Macros can't work with commands that initiate palettes or dialog boxes like LAYER or INSERT. So, we will use the command prompt version of these by adding a hyphen before the command like -LAYER and -INSERT.
- Before you start writing a macro, test the command prompts using Auto-CAD prompts that appear at the Command window. Let's do the following:
    - Type "circle" at the Command window, then press [Enter].
    - See the prompts. What is the default option? The answer is "Center."
    - For center we will use \.
    - The next prompt will be to input either the radius or diameter. With radius as the default, we will input value = 2.
- The macro will look like the following:

$$^C^Ccircle;\2;$$

- To add a new command, take the following steps:
    - Start the CUI command.
    - Using the lower left (the Command List) click Create a new command button:

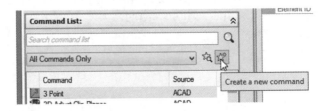

- A new command called command1 will be added.
- A Properties window will open at the right.
- Type the new name of the command (in our case we will call it CircleR2).
- Input the macro you created.
- You will see the following:

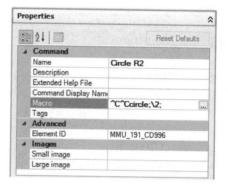

- At the top right, you will see a section called Button Image. You can select an existing image and use it as is, or select it and make some modifications. Click the Edit button, and you will see the following dialog box:

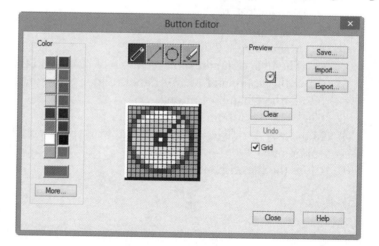

- All of the above are very familiar functions like adding a line, circle, or erasing. Use the Color section to set the current color to be used, and Grid to show rows and columns to make it easier for you to draw.

- When done, click Save, and give the new image a name.
- Accordingly the new image will be used as a representation of the new macro you just created.
- Copy it to an existing panel to be used.

PRACTICE 18-3

 Creating New Commands

1. Start AutoCAD 2014.
2. Start a new file using **My Company.dwt**.
3. Start the CUI command.
4. Create a new command with the following specifications:
   a. Name = Circle R2
   b. Description = This command will draw a circle with R = 2 units
   c. Macro = ^C^Ccircle;\2;
   d. The image should show a circle with number 2 inside it.
5. Copy the command to My Panel, under Sub-Panel 1, under Row 1, beside the Rectangular Array button.
6. Create another command with the following specifications:
   a. Name = CL 0
   b. Description = This command will make 0 the current layer
   c. Macro = ^C^C-layer;m;0;;
   d. The image should show a couple of layers and then 0 (zero).
7. Copy the command to My Panel, under Sub-Panel 1, under Row 2, beside the Chamfer button.
8. Click OK to end the CUI command.
9. Make My Workspace the current workspace and test the two new commands.
10. Close the file without saving.

NOTES:

CHAPTER REVIEW

1. Creating a template file involves which of the following?
   a. Creating text styles
   b. Setting up units and limits
   c. Creating layers
   d. All of the above
2. _____ is the command used to customize the interface.
3. You can't save your template file except in the Template folder designated by AutoCAD.
   a. True
   b. False
4. While customizing the AutoCAD interface:
   a. You can create a new workspace.
   b. You can create a new panel.
   c. You can create new commands using macros.
   d. All of the above.
5. Macros can't work with commands that show dialog boxes.
   a. True
   b. False
6. Which of the following simulates [Enter] in a macro?
   a. Backslash (\)
   b. Semicolon (;)
   c. ^C
   d. ^C^C

CHAPTER REVIEW ANSWERS

1. d
3. b
5. a

19 PARAMETRIC CONSTRAINTS

Chapter

In This Chapter

◇ What are parametric constraints?
◇ Geometric parametric constraints
◇ Dimensional parametric constraints

19.1 WHAT ARE PARAMETRIC CONSTRAINTS?

- Design software such as Inventor has used the concept of parametric constraints for a long time. Now, AutoCAD includes this concept, so we can now consider AutoCAD a designing tool as well as a drafting tool.

- Parametric constraints are geometric and dimensional. The first type will create a relationship between different objects such as parallel, horizontal, concentric, etc. A dimensional constraint will impose a certain dimension on a line, radius to an arc, or a circle, and can also create a relationship between two or more objects by writing a formula.

- With this in hand, designers have the ability to better express design intentions and protect their designs as well since the constraints make sure objects stay as they are.

- This is a huge step forward for AutoCAD and designers.

- In the following sections, we will discuss the two types of constraints and how they are applied to objects.

19.2 USING GEOMETRIC CONSTRAINTS

- Geometric constraints help you set rules for objects. You can decide to make a line always horizontal and perpendicular to another line. You can also set two circles to always share the same center point, etc. In order to use to this type of constraint, select the **Parametric** tab and locate the **Geometric** panel:

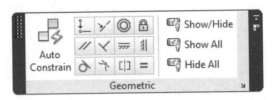

- This panel shows the 12 different types of constraints and other buttons such as the Auto Constraint button and the Show/Hide buttons. We will discuss each of these in the following sections.

19.2.1 Using the Coincident Constraint

- Locate the **Geometric** panel, then select the **Coincident** button:

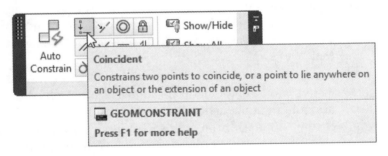

- According to AutoCAD, this constrain will constraint two points to coincide or a point to lie anywhere on an object or the extension of an object.
- You will see the following prompts:

```
Select first point or [Object/Autoconstrain] <Object>:
Select second point or [Object] <Object>:
```

- This command requires you to select two points on two existing objects. The first object will stay in its place, but the second object will move to connect with the first object. A small blue square will appear at the point connecting the two objects:

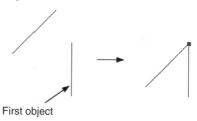

First object

- Another variation of this command is to select an object rather than a point on an object. At the first prompt, right-click and select the **Object** option, and you will see a prompt asking you to select the desired object. Once you select it, the following prompt will be shown:

```
Select point or [Multiple]:
```

- If you select a point on another object, the first object will stay in its place and the other will link with the extension of the first object just like the following:

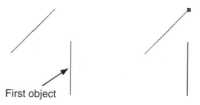

First object

- If you use the **Multiple** option, you will see the following prompts:

```
Select Point:
Select Point:
```

- You will be able to select multiple points to link them with the first object using the same rule above. The following example shows three objects coincident with the first object connected using their end points (you can select any other desired point):

19.2.2 Using the Collinear Constraint

- Locate the **Geometric** panel, then select the **Collinear** button:

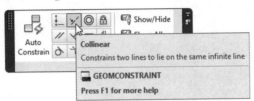

- According to AutoCAD this constraint will constrain two lines to lie on the same infinite line.
- You will see the following prompts:

```
Select first object or [Multiple]:
Select second object:
```

- The simplest method is to select the first line (which will stay intact), then select the second line that will move to be collinear. See the following illustration:

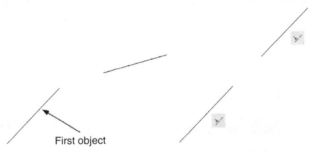

First object

- If you select the **Multiple** option, you can set several lines to be collinear.

19.2.3 Using the Concentric Constraint

- Locate the **Geometric** panel, then select the **Concentric** button:

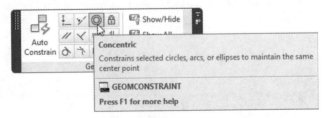

- According to AutoCAD this constraint will constrain selected circles, arcs, or ellipses to have the same center point.
- You will see the following prompts:

```
Select first object:
Select second object:
```

- See the following illustration:

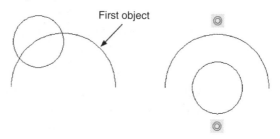

19.2.4 Using the Fix Constraint

- Locate the **Geometric** panel, then select the **Fix** button:

- According to AutoCAD this constraint will constrain a point or a curve to a fixed location and orientation relative to the World Coordinate System. You will see the following prompt:

```
Select point or [Object] <Object>:
```

- Either you can select a point on an object or select the **Object** option to select the desired object. See the following illustration:

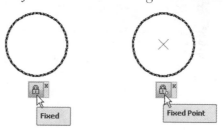

19.2.5 Using the Parallel Constraint

- Locate the **Geometric** panel, then select the **Parallel** button:

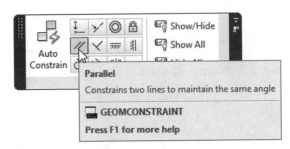

- According to AutoCAD this constraint will constrain two lines to be parallel. You will see the following prompts:

```
Select first object:
Select second object:
```

- The first object will keep its current angle, but the second object will rotate to have the same angle as the first. See the following illustration:

19.2.6 Using the Perpendicular Constraint

- Locate the **Geometric** panel, then select the **Perpendicular** button:

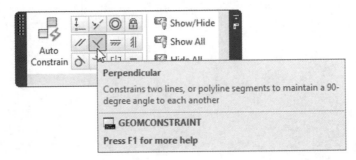

■ According to AutoCAD this constraint will constrain two lines or polyline segments to maintain a 90-degree angle to each other. You will see
the following prompts:

```
Select first object:
Select second object:
```

■ The first object will keep its current angle, and the second object will
rotate to make it 90 degrees relative to the first object. See the following
illustration:

19.2.7 Using the Horizontal Constraint

■ Locate the **Geometric** panel, then select the **Horizontal** button:

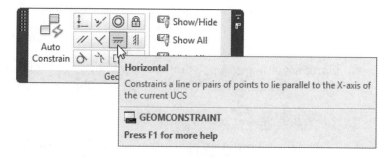

■ According to AutoCAD this constraint will constrain a line or pair of
points to lie parallel to the x-axis of the current UCS. You will see the
following prompt:

```
Select an object or [2Points] <2Points>:
```

■ If you select a line it will become horizontal and the command will end.
The endpoint nearest to the selection point will remain, but the other
end will move.

- See the following illustration:

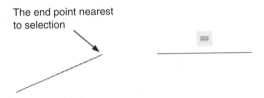

The end point nearest
to selection

- If you select the **2Points** option, you will see the following prompts:

```
Select first point:
Select second point:
```

- This option will make sure that the two points selected will form an imaginary horizontal line. See the following example:

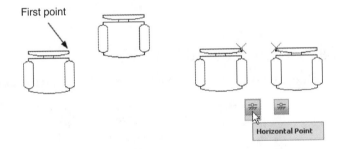

First point

Horizontal Point

- Note that although a polyline is considered a single object, if you use this constraint on it, AutoCAD will treat each object separately.

19.2.8 Using the Vertical Constraint

- Locate the **Geometric** panel, then select the **Vertical** button:

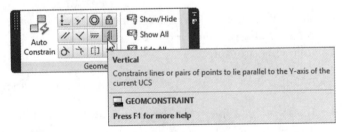

Vertical
Constrains lines or pairs of points to lie parallel to the Y-axis of the current UCS

GEOMCONSTRAINT
Press F1 for more help

- This command is identical to the above command and allows objects or points to be vertical.

19.2.9 Using the Tangent Constraint

- Locate the **Geometric** panel, then select the **Tangent** button:

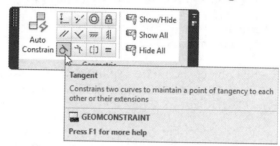

- According to AutoCAD this constraint will constrain two curves to maintain a point of tangency to each other, or their extension. You will see the following prompts:

```
Select first object:
Select second object:
```

- The first object will stay in its current place, but the second object will move using the nearest tangent point to its current position. See the following illustration:

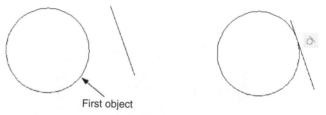

First object

19.2.10 Using the Smooth Constraint

- Locate the **Geometric** panel, then select the **Smooth** button:

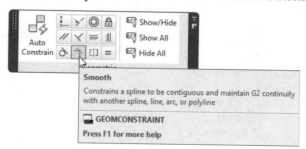

- According to AutoCAD this constraint will constrain a spline to be contiguous and maintain G2 continuity with another spline, line, arc, or polyline. You will see the following prompts:

```
Select first spline curve:
Select second curve:
```

- The first object should be a spline, but the second object can be a spline, line, arc, or polyline. See the following illustration:

- Note that although we are selecting objects, the point we highlight during the selecting is very important to the end result.

19.2.11 Using the Symmetric Constraint

- Locate the **Geometric** panel, then select the **Symmetric** button:

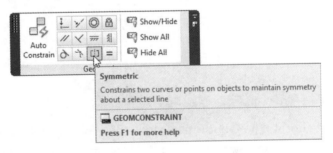

- According to AutoCAD this constraint will constrain two curves or points on objects to maintain symmetry about a selected line. You will see the following prompts:

```
Select first object or [2Points] <2Points>:
Select second object:
Select symmetry line:
```

- This is similar to the Horizontal command, because it uses the same prompts. You will select either two objects, or two points. The first object will maintain its current angle, but the second object rotate to make the mirror image of the first object around the symmetry line selected. See the following illustration:

19.2.12 Using the Equal Constraint

- Locate the **Geometric** panel, then select the **Equal** button:

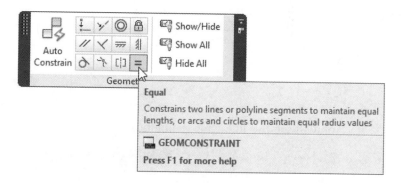

- According to AutoCAD this constraint will constrain two lines or polyline segments to maintain equal lengths, or for arcs and circles to maintain equal radius values. You will see the following prompts:

```
Select first object or [Multiple]:
Select second object:
```

- The first object will maintain its length (radius for arcs and circles), and the second object will change its length to match the first object. If you select the **Multiple** option, you can match the length of the first object with multiple lines rather than with a single line. See the following illustration:

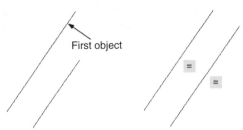

First object

19.3 GEOMETRIC CONSTRAINTS SETTINGS

- To control the display of the geometric constraints, you should use the **Settings** dialog box. To issue this command, locate the **Geometric** panel, then click the arrow at the lower-right corner of the panel:

- You will see the following dialog box:

- You will see that by default all of the constraints will be displayed, but you can select to hide any of the 12 constraints in a drawing. You can also control the following:
  - Whether to only display constraint bar for objects in the current plane
  - The constraint bar transparency value
  - Whether or not to show the constraint bar after applying a constraint to selected objects
  - Whether or not to show the constraint bar when objects are selected

19.4 WHAT IS THE INFER CONSTRAINT?

- In the above dialog box, one checkbox was overlooked, the Infer geometric constraint checkbox, at the top left of the dialog box. So, what is the Infer constraint? This constraint flips the process of constraining in AutoCAD, as it allows you to set the constraints while drafting rather than after drafting. Commands like Fillet and Chamfer will be affected.
- To activate this command, go to the status bar and click the **Infer Constraints** button:

- After you switch this button on, AutoCAD will apply constraints to all objects as they are added to the drawing, using proper constraints depending on the object and the objects attached to it. While this method isn't a replacement for the first method discussed, will help you complete your work in fewer steps. Let's take a look at the following example:
  - Switch on the Infer Constraint button.
  - Start the Line command and draw a vertical line. You will notice that AutoCAD applies a vertical constraint.
  - Continue by drawing a horizontal line. You will notice that AutoCAD applies a perpendicular constraint to the new horizontal line.
  - Continue drawing another vertical line, close the rectangle, and you will notice that AutoCAD applies the perpendicular constraint, along with coincident on all corner points.

- Start the Fillet command and set the radius to a suitable value, then fillet one of the corners. You will notice that AutoCAD applies the tangent constraint between the two lines and the added arc, along with the coincident constraint at the two ends of the arc.
- You will see something like the following:

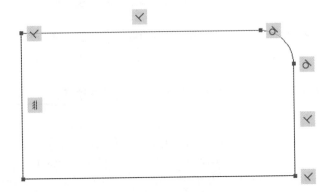

- If you right-click the button at the status bar, and select **Settings**, you will see the same dialog box we saw in the previous section, which will control which constraint to be displayed on the screen and which is not.

19.5 WHAT IS AUTOCONSTRAIN?

- The AutoConstrain command allows you to apply multiple geometric constraints to selected objects, based on the current relationship between these objects and the selected constraint to be applied in the Settings dialog box of the auto-constraint.
- To issue this command, locate the **Geometric** panel, and then select the **AutoConstrain** button:

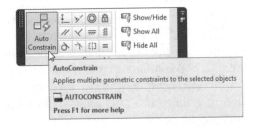

- You will see the following prompt:

```
Select objects or [Settings]:
```

- As a first step, select the **Settings** option, and you will see the following dialog box:

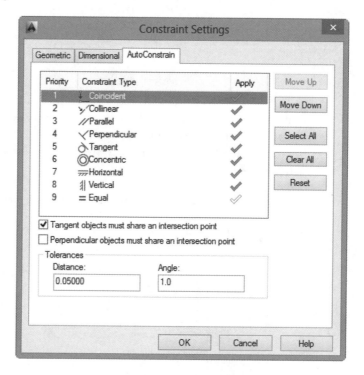

- Using this dialog box you can do all or any of the following:
  - Move the constraints up and down to set the priority for each constraint
  - Turn off any unwanted constraint by clicking the green (✓)
  - Use the Select All, Clear All, and Reset buttons
  - Decide whether tangent objects must share an intersection point or not
  - Decide whether perpendicular objects must share an intersection point or not
- As you can see three of the normal 12 constraints will not be used: Symmetrical, Fix, and Smooth.

■ Let's look at an example. You have the following shape:

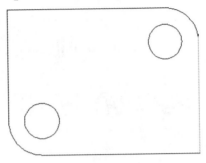

• Before you apply any constraint, select all the object using grips. Try to move one of the grips, and make a note of its movement.
• Start the **AutoConstrain** command.
• Select all the objects, then press [Enter], and the following message will appear:

```
16 constraint(s) applied to 8 object(s)
```

• The picture will change to:

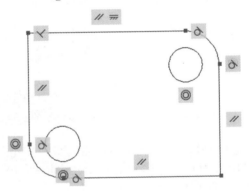

• Select the objects using grips, and try to move one of the grips. You will notice that all the objects move together as a coherent set of objects that understand each other and keep the correct relationships at all times.
■ As a final note on AutoConstrain, note that the first prompt of the coincident constraint includes AutoConstrain:

```
Select first point or [Object/Autoconstrain] <Object>:
```

■ This means if you select this option and then select objects, all of them will have a coincident constraint.

19.6 CONSTRAINT BAR AND SHOW & HIDE

19.6.1 Constraint Bar

- The constraint bar is a small bar that appears beside an object after you apply a geometric constraint to it. Hovering over the small bar will highlight the objects affected. See the following illustration:

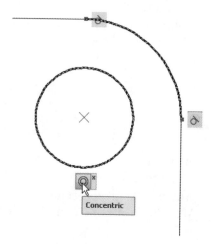

- On the other hand, if you hover over an object with a geometric constraint applied to it, the constraint bar will highlight as well. Occasionally one object may hold more than one constraint, so the bar shows more buttons.
- Right-clicking the buttons on the constraint bar shows the following menu:

- In this menu, you can:
  - Delete the selected geometric constraint; you can also use the [Del] key on the keyboard
  - Hide the current Constraint Bar (even if it contains more than one button)

- Hide All Constraints in the current drawing
- Show the Constraint Bar Settings, which was discussed previously
- You can move the constraint bar from its default position to any other position. Simply click and hold and drag it to a new location.

19.6.2 Show & Hide

- You can also control the visibility of the constraint bar. You can control whether to show it or not for all objects or just for some objects.
- To issue this set of commands, locate the **Geometric** panel to issue one of the following commands:

- The commands are:
  - Show/Hide: to hide some of the constraints and show others
  - Show All: to show all constraints for all objects
  - Hide All: to hide all constraints for all objects

19.7 RELAXING AND OVER-CONSTRAINING OBJECTS

19.7.1 Relaxing Constraints

- Relaxing an object means removing some of the constraints applied to it, in order to fulfill a command that wants to change the object's status. Assume you have a line with a parallel constraint to a vertical line and you try to rotate it. How will AutoCAD respond? You will see the following message:

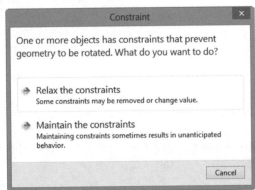

- As you can see, AutoCAD will give you the choice either to relax the constraints or to maintain them. Selecting relaxing will remove only the associated constraint that allows the command to be fulfilled.

19.7.2 Over-Constraining an Object

- Let's assume you applied a horizontal constraint to a line and then you applied the perpendicular constraint to an attached line. Applying vertical to the second line will mean this object is over-constrained, and Auto-CAD will not allow this to happen. You will see the following message:

- AutoCAD will not allow this action, so you will click OK or Cancel to abort the process.

PRACTICE 19-1

Applying Geometric Constraints

1. Start AutoCAD 2014.
2. Open **Practice 19-1.dwg**.
3. Make sure that the Infer Constraint button is off at the status bar.
4. Start coincident, select AutoConstrain, and then select the whole shape.
5. Start concentric, and select the arc, then the circle.
6. Start vertical, and select the right vertical line.
7. Start perpendicular, and select the right vertical line and the upper horizontal line.
8. Repeat the same with the lower horizontal line.
9. Start parallel, and select the right vertical line and the left vertical line.
10. Start horizontal, and select the lower horizontal line. What message appears? _____

11. Click on the Infer Constraint button at the status bar.
12. Using the Fillet command, set radius = 2, and fillet the upper-right corner of the shape.
13. Draw a circle with radius = 1, using the center of the arc you just added using the Fillet command.
14. Select the whole shape using grips, click the mid-point of the right vertical line and move it, and see how the whole shape responds.
15. Switch off the Infer Constraint button.
16. Pan to the shape at the right.
17. Start the AutoConstrain command, select Settings, and click the Select All button to select all constraints.
18. Click OK, and select the whole shape. How many constraints were added? _____ (18) On how many objects? _____
19. Notice the equal constraints for both circles and arcs, delete them.
20. Select to Hide all constraints, then select to show them only for some objects.
21. Save and close the file.

19.8 USING DIMENSIONAL CONSTRAINTS

- While geometric constraints are useful, we also need to set dimensional constraints to use the constraint concept to its fullest. Dimensional constraints allow you to specify a length for a line, an angle between two lines, a radius, or a diameter for an arc, or a circle. You can also link the different dimensional constraints using formulas.
- To reach to all the dimensional constraints commands, select the **Parametric** tab and then locate the **Dimensional** panel:

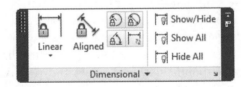

- The following sections cover all the types of dimensional constraints.

19.8.1 Using Linear, Horizontal, and Vertical Constraints

- These constraints deal with distances between points. If you click the button, you will see the following options:

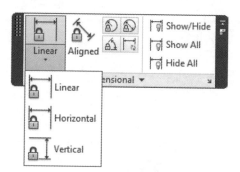

- According to AutoCAD, linear will constrain either horizontal or vertical lines between two points (even if the two points form an angle other than 0, 90, 180, or 270). The Horizontal command will constrain the x-axis distance between two points, and Vertical will constrain the y-axis distance between two points. You will see the following prompts:

```
Specify first constraint point or [Object] <Object>:
Specify second constraint point:
Specify dimension line location:
Dimension text = 6.6615
```

- The prompts for the three commands are the same. If you select the first point (a red cross with a circle will appear) and then you select the second point. You will be asked to specify the dimension line location (just as we did in dimensioning) and then whether to accept the real measured distance by pressing [Enter] or to input your own value. Of course, the length will change according to the new value. If, at the first prompt, you selected the Object option, you will see the following prompts:

```
Select object:
Specify dimension line location:
Dimension text = 9.7586
```

- The following illustration shows the shape of the dimensional constraint:

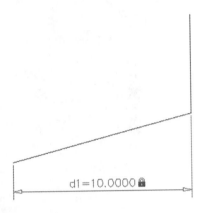

19.8.2 Using Aligned Constraints

- To issue this command, locate the **Dimensional** panel and then select the **Aligned** button:

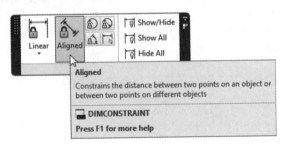

- According to AutoCAD, this constraint will constrain the distance between two points whether on the same objects or on different objects. You will see the following prompt:

```
Specify first constraint point or [Object/Point &
line/2Lines] <Object>:
```

- If you specify the first constraint point, you will be asked to specify the second point. If you use the Object option, you will be asked to select an object. If you chose the Point & line option you will see the following prompts:

```
Specify constraint point or [Line] <Line>:
Select line:
```

```
Specify dimension line location:
Dimension text = 2.2466
```

- You will select a point, then a line (while selecting a line, make sure you are selecting the right point). If you select the 2Lines option, you will see the following prompts:

```
Select first line:
Select second line to make parallel:
Specify dimension line location:
Dimension text = 2.9135
```

- Selects two line objects. The second line will be made parallel to the first line. The aligned constraint controls the distance between the two lines.
- The following illustration shows the aligned constraint:

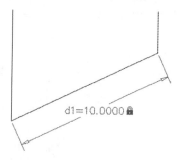

19.8.3 Using Radial and Diameter Constraints

- To issue this command, locate the **Dimensional** panel, and then select the **Radial** or **Diameter** button:

- According to AutoCAD, these two constraints will constrain the radius or the diameter of a circle or arc. You will see the following prompts:

```
Select arc or circle:
Dimension text = 3.4359
Specify dimension line location:
```

■ The following illustration shows the radial and diameter constraints:

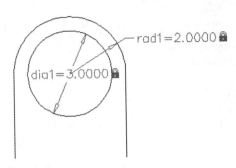

19.8.4 Using Angular Constraints

■ To issue this command, locate the **Dimensional** panel, and then select the **Angular** button:

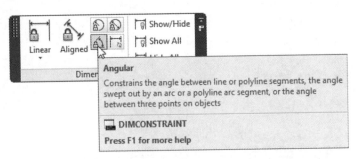

■ According to AutoCAD this constraint will constrain the angle between two lines or polyline objects, in an arc, or using three points. You will see the following prompts:

```
Select first line or arc or [3Point] <3Point>:
Select second line:
Specify dimension line location:
Dimension text = 45
```

■ You will select two lines (or polyline segments) or an arc. If you select the 3Point option, you will see the following prompts:

```
Specify angle vertex:
Specify first angle constraint point:
Specify second angle constraint point:
```

- You will select the vertex point first, then the two other points to form an angle. The following illustration shows the angular constraint:

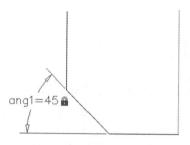

19.8.5 Using the Convert Command

- To issue this command, locate the **Dimensional** panel, and then select the **Convert** button:

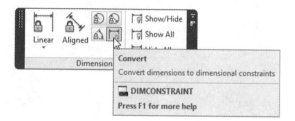

- This command will convert normal dimensions to dimensional constraints. You will see the following prompts:

```
Select associative dimensions to convert:
Select associative dimensions to convert:
```

- Simply select the desired dimensions to be converted to constraints. The following illustration shows the process:

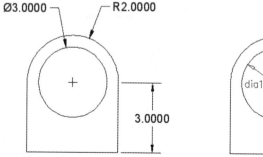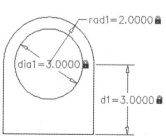

19.9 CONTROLLING DIMENSIONAL CONSTRAINTS

- In this section, we will discuss how to control dimensional constraints by:
  - Using the Constraint Settings dialog box
  - Deleting Dimensional constraints
  - Showing and hiding dimensional constraints

19.9.1 Constraint Settings Dialog Box

- In this dialog box, you will control the appearance of the dimensional constraint on the screen. To issue this command, select the **Parametric** tab, locate the **Dimensional** panel, and then select the **Constraint Settings, Dimensional** button:

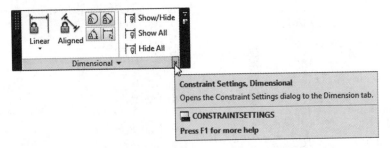

- The following dialog box will appear:

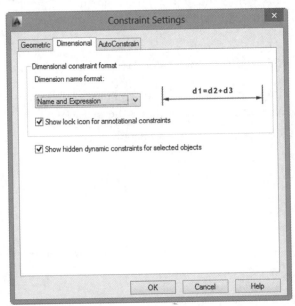

- This dialog box allows you to control what will appear at the dimensional constraint; there are three choices:
  - Name
  - Value
  - Name and Expression
- You can select whether to show or to hide the lock symbol that appears near the measured value. You can also select whether to show any hidden dimensional constraints when object you select an object. See the following example:

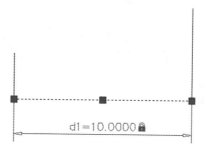

- In the above, the dimensional constraint was hidden. When we selected the object, the dimensional constraint appeared.

19.9.2 Deleting Constraints

- You can use the [Del] key on the keyboard to get rid of any dimensional or geometric constraint. You can also locate the **Manage** panel and then select the **Delete Constraints** button:

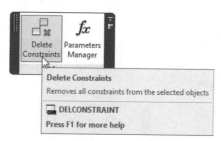

- The following prompt will appear:

```
All constraints will be removed from selected objects…
Select objects:
```

- Keep selecting constraints, and press [Enter] to end the command when you are done.

19.9.3 Showing and Hiding Dimensional Constraints

- These commands are similar to what we learned for geometric constraints. To issue this command, locate the **Dimensional** panel, and then select one of the following three buttons:

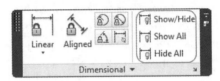

- The Show/Hide button shows a hidden constraint or hides a visible constraint by selecting. You will see the following prompts:

```
Select objects:
Select objects:
Enter an option [Show/Hide]<Show>:
```

- The Show All button shows all dimensional constraints.
- The Hide All button will hide all dimensional constraints.

19.10 USING THE PARAMETERS MANAGER

- The Parameters Manager is the place to create equations involving dimensional constraints, i.e., where you will link dimensional constraints together. Using this method any change in one of the dimensional constraints will affect the other. In order to create the right equations, AutoCAD allows us to create user-defined parameters. To issue this command, locate the **Manage** panel, and then select the **Parameters Manager** button:

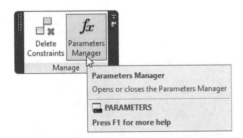

- You will see something like the following palette:

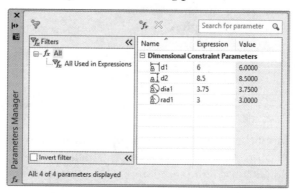

- The Parameters Manager palette will be displayed and shows all the existing dimensional constraint parameters. As you can see, AutoCAD uses the default names, but you can also rename these parameters. Simply click the name once, and you will be able to input your own name:

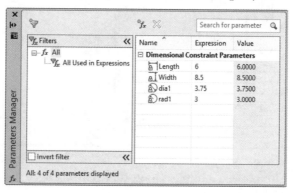

- And you can create an equation including one or more of the parameters as in the following example:

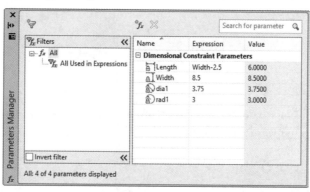

- Whenever you change Length, the Width will change as well. You can create your own parameters by creating user-defined parameters, using the following button:

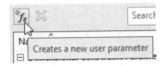

- Input the name of the parameter and an expression (if valid); if not, input the current value. Later, you can include an expression involving dimensional constraint parameters. Look at the example below:

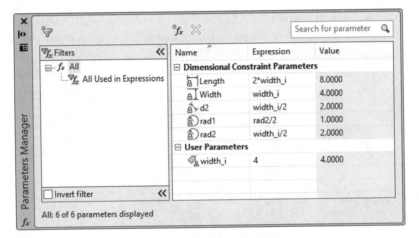

- The above example shows that both Length and Width are now linked with Step_Length.
- AutoCAD also provides filters to categorize your parameters in categories you create. To create a new filter, click the following button:

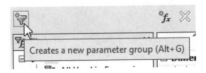

- Once you click this button, a new filter will be added. Simply type the name of the filter, then press [Enter]. To fill it with parameters, drag parameters from the right and drop them at the name of the filter. You will get something like the following:

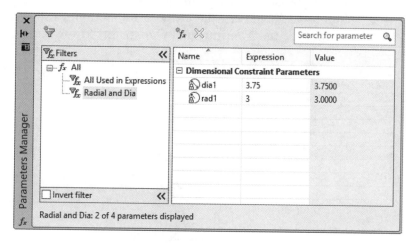

- You can easily differentiate between parameters and equations, equations starts with **fx:**. You can overwrite an equation by double-clicking it and inputting a value.

19.11 WHAT IS ANNOTATIONAL CONSTRAINT MODE?

- The blocks displayed after you add a dimensional constraint will not be printed, and zoom in and zoom out commands will not affect displayed size. This means you will have two things to do. You will add normal dimensions and will also have to add dimensional constraints! But Auto-CAD can help. You can convert all the dimensional constraints to annotational constraints. Annotational constraints are printed and use the current dimension style.
- There are two ways to do this:
  - Before you start, you can change the mode to use annotational constraints. With this method, you will combine the two actions into a single action.
  - For the dimensional constraints already placed, you can convert them using the Properties palette.

19.11.1 Annotational Constraint Mode

- Before you start placing dimensional constraints, locate the **Dimensional** panel and click the arrow at the bottom to show more buttons, making sure the **Annotational Constraint Mode** is selected:

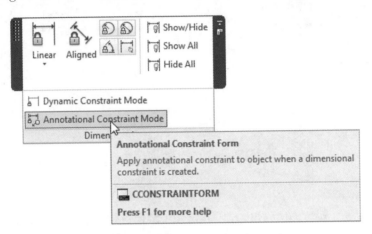

- Now add a dimensional constraint using the ways we discussed previously, and you will see something like the following:

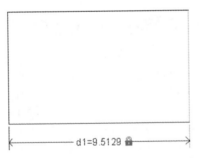

- Using the Dimensional Constraint Settings dialog box, you can choose to show/hide the lock symbol along with showing the name.

19.11.2 Converting Dimensional Constraints to Annotational

- If you already placed dimensional constraints, you can convert them to be annotational constraints using the Properties palette. Simply select the desired dimensional constraint to be converted, right-click, and select Properties. You will see the following palette; click **Constraint Form** and select **Annotational** instead of **Dynamic**:

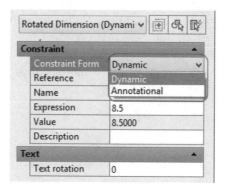

19.12 USING DIMENSIONAL GRIPS

- You can manipulate dimensional constraints using grips. The image below includes three types of dimensional constraints: linear, aligned, and angular:

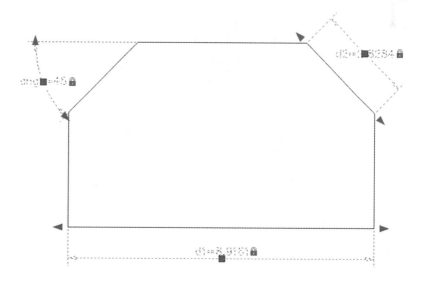

- As you can see there are grips at the middle of each dimensional constraint that allow you to move the whole block back and forth. The other two arrows allow you to stretch the dimensional constraint to stretch the object by itself. Dimensional parameters have their own grips depending on the type.

■ In the following example, there are radius and diameter constraints. The grip will help you relocate the dimensional constraint block. While radius has one arrow to change the value of the radius, the diameter has two arrows to change the value in either direction. Arcs and circles will be affected by this change:

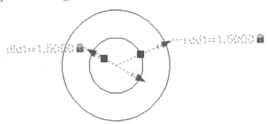

PRACTICE 19-2

   Applying Dimensional Constraints

1. Start AutoCAD 2014.
2. Open **Practice 19-2.dwg**.
3. Using different types of dimensional constraints add the following constraints using the same names:

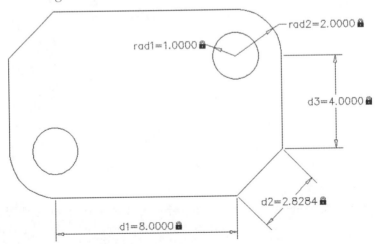

4. Using the Parameters Manager do the following:
   a. Rename d1 as Length
   b. Rename d3 as Width
   c. Create a user-defined parameter width_i with a default value of 4
   d. Change Width = width_i
   e. Change Length = 2 * width_i
   f. rad2 = width_i / 2
   g. d2 = width_i / 2
   h. rad1 = rad2/2
5. Using the Parameters Manager change the values of width_i to the following values: 2, 3, 5, 6. How do these changes affect the other objects?

   _____

   _____

   _____

6. Change width_i to 4, and close the Parameters Manager.
7. Change one of the dimensional constraints to be annotational using zoom in and zoom out. How does this block respond compared to the other blocks?
8. Using the Dimensional Constraint Settings, choose to show only the values without the lock symbol.
9. The lock symbol is still there because all the constraints are equations.
10. Save and close the file.

NOTES:

CHAPTER REVIEW

1. Which one of the following is *not* related to dimensional constraints?
   a. They are not affected by zooming
   b. Linear
   c. Equal
   d. Angular
2. Parallel, fix, and coincident are all _____ constraints.
3. Infer constraints can be applied while:
   a. Adding a line command
   b. Filleting two lines
   c. Stretching objects
   d. Chamfering two lines
4. You can add a single block as a normal dimension and as a constraint.
   a. True
   b. False
5. There is an option called AutoConstrain in the Collinear command.
   a. True
   b. False
6. You have to select a spline as the first object for the _____ constraint.
7. In _____, you can create equations linking different dimensional constraints together.

CHAPTER REVIEW ANSWERS

1. c
3. c
5. b
7. Parameters Manager

Chapter **20** # DYNAMIC BLOCKS

In This Chapter
◇ What are dynamic blocks?
◇ Parameters and actions
◇ Dynamic blocks and constraints

20.1 INTRODUCTION TO DYNAMIC BLOCKS

- We know that AutoCAD can create and insert a block with one shape and dimension that can be scaled bigger or smaller. But AutoCAD also provides the ability to create a block with multiple shapes, sizes, and views. This type of block is called a dynamic block and can be created in the Block Editor.
- Dynamic blocks use parameters and actions. Actions are very similar to AutoCAD modifying commands, such as Stretch, Scale, and Array. Each action needs a parameter to work on. You can also create dynamic blocks using geometric and dimensional constraints (discussed in Chapter 14).

■ Dynamic blocks always have a lightning symbol in its right corner, whether you are at the Insert command dialog box or in tool palettes. Here are some examples:

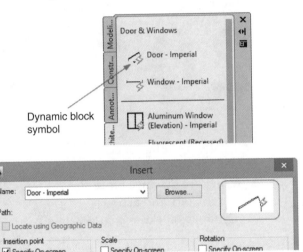

■ Dynamic blocks are very handy tools to help you improve your productivity and reduce the number of blocks you need in a drawing.

20.2 HOW TO CREATE A DYNAMIC BLOCK

■ Dynamic blocks are made in Block Editor. You can reach Block Editor several ways:
   • Using the **Open in block editor** checkbox in the Block Definition dialog box
   • By double-clicking an existing block, then selecting the name of the block
   • By issuing the **Block Editor** command

20.2.1 Using the Block Definition Dialog Box

- In general, when you create a normal block, you will use the Block Definition dialog box as shown below. Simply click the Open in block editor checkbox on, and then click OK, and AutoCAD will take you directly to the Block Editor where you can add dynamic features to your normal block:

20.2.2 Double-Clicking an Existing Block

- You can convert an existing normal block to a dynamic by double-clicking it. You will see the following dialog box:

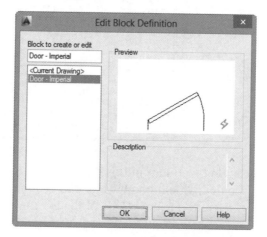

- AutoCAD shows all the blocks in the current drawing. Select the block you want to change and then click OK to go directly to the Block Editor to add features.

20.2.3 Block Editor Command

- To issue this command, go to the **Insert** tab and the **Block Definition** panel, then select the **Block Editor** button:

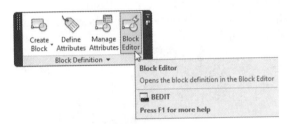

- The following dialog box will appear:

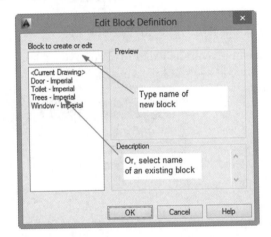

- You can do two things in this dialog box:
  - Type the name for a new block to be created in the Block Editor
  - Or, select the name of one of the existing blocks to convert (or edit)

20.3 INSIDE BLOCK EDITOR

- Once you are in the Block Editor, three things will happen:
  - The background will change to a different color (most likely it will be dark grey).
  - A new contextual tab called **Block Editor** will appear, which includes several panels.

- The **Block Authoring** palette will appear, which contains four tabs: Parameters, Actions, Parameter Sets, and Constraints, just like the following:

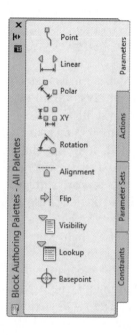

- We will use both the context tab "Block Editor" panels and the Block Authoring palettes to add dynamic features to our block.

20.4 WHAT ARE PARAMETERS AND ACTIONS?

- There is a very simple rule for creating dynamic block. First, you add a parameter to your block and then assign an action to work on it. Parameters are geometric features of the objects forming the block, such as linear, polar, visibility, lookup, etc. Actions are a simulation of AutoCAD modifying commands such as Move, Stretch, Scale, etc.
- You will find that some parameters don't need actions, and some actions don't need parameters to work on. But the rule-of-thumb is we need them in this order: parameter first, then action second.

AutoCAD includes parameter sets, which include a set of parameters with actions.

- Both parameters and actions have properties, and these properties control the outcome of the dynamic block. In Chapter 19, we discussed geometric and dimensional constraints and how they affect normal objects. AutoCAD gives us the ability to enhance these constraints inside a block.

- The following lists the available parameters:
  - Point parameters will add a point definition, and stretch and move actions will use it.
  - Linear parameters measure a distance, which can have any angle. But once applied, you can't change it. These actions act on these parameters: move, stretch, scale, and array.
  - Polar parameters are similar to linear parameters and have the ability to change the angle. These actions can act on these parameters: move, stretch, polar stretch, scale, and array.
  - XY parameters help you define two associated dimensions in the X and Y directions. These actions can act on these parameters: move, stretch, scale, and array.
  - Rotate parameters help you add a rotation parameter to the block definition. The rotate action can act on this parameter only.
  - Alignment parameters help you add the ability for the block to align itself with an existing object. This parameter doesn't need an action.
  - Flip parameters allow you to flip an existing block to its mirror image using the mirror line. The flip action can act on this parameter only.
  - Visibility parameters help you create visibility state for some of the objects to be visible and others to be invisible. This parameter doesn't need an action.
  - Lookup parameters help you create a table of possible values of existing parameters. The lookup action can act on this parameter only.
  - Basepoint parameters define a new basepoint for the block. This parameter doesn't need an action.

- The following is a list of the available actions:
  - The move action will move the selected objects from one position to another. You will see the following prompts:

```
Select parameter:
Specify selection set for action
Select objects:
```

  - The scale action will scale the selected objects. You will see the following prompts:

```
Select parameter:
Specify selection set for action
Select objects:
```

  - The stretch action will stretch the selected objects using the crossing concept (just like in the normal Stretch command). You will see the following prompts:

```
Select parameter:
Specify parameter point to associate with action or
enter [sTart point/Second point] <Second>:
Specify first corner of stretch frame or [CPolygon]:
Specify objects to stretch
Select objects:
```

  - The polar stretch action is equal to the stretch action, but it has an angle option.
  - The rotate action will rotate objects using the rotate parameter. You will see the following prompts:

```
Select parameter:
Specify selection set for action
Select objects:
```

- The flip action will work on the flip parameter only and will create a mirror image using the mirror line created with the flip parameter. You will see the following prompts:

```
Select parameter:
Specify selection set for action
Select objects:
```

- The array action simulates the Rectangular Array command (the old array before the 2012 version). You will see the following prompts:

```
Select parameter:
Specify selection set for action
Select objects:
Enter the distance between columns (| | |):
```

- The lookup action will work only on the lookup parameter in order to create a table of existing parameters. AutoCAD; shows them as a list for you to choose from.
- The Block Properties table will be discussed once we discuss geometric and dimensional constraints and how to include them in a dynamic block.

- Parameter sets reduce the steps needed to complete a drawing, but there are two exceptions:
  - Pairs, which means there are two actions acting on a single parameter. There are several pairs such as linear stretch pair, linear stretch pair, etc.
  - Box set, which means there are four actions on a single parameter. There are several box sets: XY stretch box set, XY array box set, etc.
- You will find the same set of parameters and actions using the context tab Block Editor, located in the **Action Parameters** panel, which looks like the following:

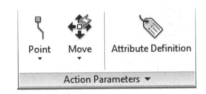

- After you finish adding parameters and actions you will see something like the following:

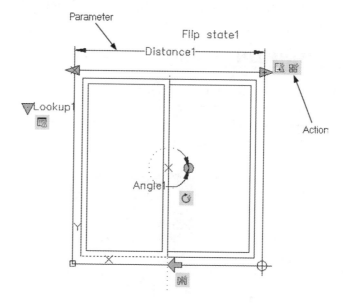

- If you hover over one of the actions, all the parameters acting on it will be highlighted, along with the objects selected:

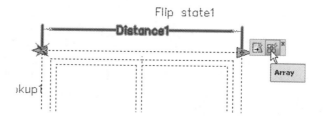

- If you right-click any action, you will see the following menu:

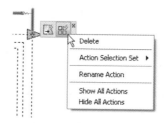

- Using this menu, you can delete the action, change the selection set, rename the actions, and show/hide all actions in the current file. This menu may show different options for different actions.

20.5 CONTROLLING PARAMETER PROPERTIES

- You can control the parameter properties to control the behavior of the block once you start to use it. To see the properties palette, select the parameter, right-click, then select the **Properties** option, and something like the following will appear:

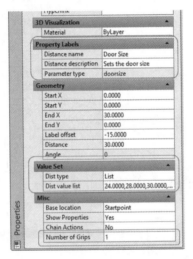

- When you add a new parameter, AutoCAD will give it a temporary name, but you can rename it to something specific.
- To do that locate the **Property Labels** and change the **Distance name** (this example is for a linear parameter), just like the following:

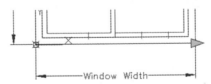

- Another thing you can control is size using **Value Set**. There are three different types:

- None: This type means you can input two values, the Dist minimum and Dist maximum. These two values can be different or can be equal:

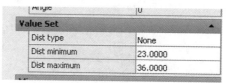

- Increment: This type means the value will change using a Dist Increment. You should input both the Dist minimum and the Dist maximum:

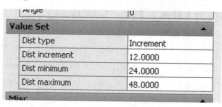

- List: This type means the value will change using a list of values that will be input using the small button at the right with the three dots inside it. When clicked, you will see the next dialog box:

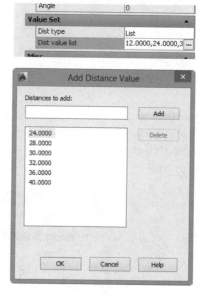

- Finally, you can control the number of grips to appear in the parameter. The default number of parameters in linear is two and in XY is four.

In order to prevent you from scaling or stretching your block from points you don't want, simply reduce or remove the grips. See the following illustration:

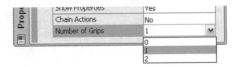

20.6 CONTROLLING THE VISIBILITY PARAMETER

- One of the features you can add to a dynamic block is visibility. The technique is very simple. Once you have several objects (blocks) you create a state, which shows some of these objects and hides others. You will create several states. To use this block, you will choose from a list of states beside the block. You will do to the following:
  - Add a visibility parameter (this parameter doesn't need an action)
  - Once you add it, the **Visibility** panel in the Block Editor context tab will come to life, just like the following:

- As you can see, AutoCAD created a new visibility state and gave it a temporary name, VisibilityState0. Click the **Visibility States** button to rename the current visibility state, and you will see the following dialog box:

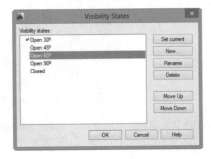

- Click the **Rename** button and input a new name, then click OK.
- Select the object(s) you want to make visible, then click the **Make Visible** button at the Visibility panel. Then select the object(s) you want to make invisible, then click the **Make Invisible** button at the Visibility panel. You will see the following two buttons:

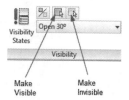

- The default is to hide 100% of the invisible objects. But you can choose to show them with a certain transparency by using the **Visibility Mode** button:

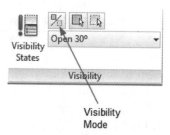

- Create another visibility state by clicking the **Visibility State** button, then clicking the **New** button. You will see the following dialog box:

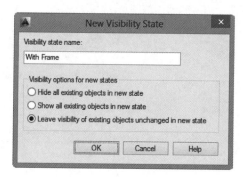

- Type the name of the new state, and then select one of the available choices. You can select to Hide/Show all of the objects or keep the objects with their current visibility settings. Click OK, then change the visibility of the objects as you wish.
- Create another state and so on.
- Test your settings by using the list in the panel, just like the following:

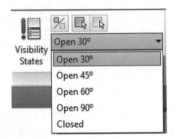

20.7 USING THE LOOKUP PARAMETER AND ACTION

- The lookup parameter and action are used to see all the different sizes and options for your block in a simple list. With this approach, you can fully control the output of the block, because you are limited to those choices you've added to the list. You should add the parameters and set the grip and value conditions, and lastly, add the lookup parameter and lookup action, or you can add both in one step using parameter sets. After adding the parameter and action take the following steps:
  - Right-click the action icon to show the following menu:

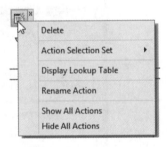

- Select the **Display lookup Table** option, and you will see the following dialog box:

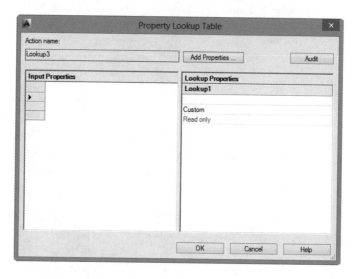

- To add the parameters you want to show in the list click the **Add Properties** button to see the following dialog box:

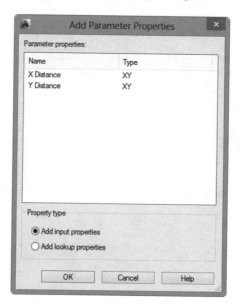

- Select the desired parameters (you can use the [Ctrl] key to select multiple parameters), then click **OK** to end the adding process.
- The left side of the dialog box will change to the following:

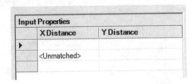

- Start selecting the different values for Window Width and the corresponding Window Height. On the right, input the text that should appear, something like "Size 1", "Size 2", etc. This is what you will have at the end:

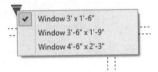

- The final product will look something like the following:

20.8 FINAL STEPS

- Once you are done adding your features, go to the outmost panel at the left, then the outmost panel at the right. The one at the left is the **Open/ Save** panel:

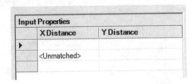

- This panel allows you to:
  - Save the block with the current addition
  - Test the block without leaving the Block Editor
  - Edit another block
  - If you expand the panel, you will see a button to Save the block under another name
- The second panel is at the rightmost panel and looks like the following:

- To finish the Block Editor command, click the **Close Block Editor** button. If you didn't save your changes, AutoCAD shows a warning message and asks you to save or discard your changes.

PRACTICE 20-1

Dynamic Blocks – Creating a Chest of Drawers

1. Start AutoCAD 2014.
2. Open **Practice 20-1.dwg**.
3. Create a new block and call it **Chest of Drawers**, setting the upper-right corner as the base point, the block units to be Centimeter, and making sure that the Open in block editor checkbox is on.
4. Add a linear parameter for the left vertical line (choose the lower point first, then the upper point). Using the Properties palette change the name to Height and set the number of grips to one.
5. Using Properties change Dist Type to Increment, with Dist minimum = 35, Dist maximum = 130, and Dist Increment = 25.
6. Add a stretch action using the following information:
   a. Select height parameter
   b. Select the upper-left corner as your parameter point

c. Set the stretch frame (and objects to select) as in the illustration below, then press [Enter]:

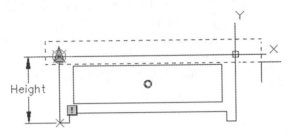

7. Add an array action, selecting the height parameter and setting the distance to be 25.
8. Without leaving Block Editor, test the block. If everything is OK, save it and test it normally.
9. Save and close the file.

PRACTICE 20-2

Dynamic Blocks – Door Control

1. Start AutoCAD 2014.
2. Open **Practice 20-2.dwg**.
3. There will be a block representing a door. Double-click the block, and a dialog box will appear. Select Interior Door (it is already selected), and click OK. You are in Block Editor.
4. Add a flip parameter by selecting the midpoint of the two jambs as the reflecting line and specifying the label location beneath the line.
5. Add a flip action by selecting the parameter and all objects representing the door.
6. Test the block.
7. Add a flip parameter. To specify the reflecting line use [Shift] + right-click, and select the Mid between Two Points option. Select the lower-left corner of the left jamb and the lower-right corner of the right jamb. Auto-CAD will select the point between these two points. Using Polar Tracking, select any point upwards or downwards. Select the label location to be at the middle of the door opening.
8. Add a flip action by selecting the parameter and all objects representing the door.

9. Test the block.
10. Draw a rectangle (using the Rectangle command or Line command) to show that the door is closed, just like the following:

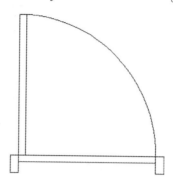

11. Add a visibility parameter, selecting the upper-left part of the door to be the location of the parameter. Note how the Visibility panel turned on.
12. Using the Visibility panel create a new state and call it Open by renaming the existing state.
13. Select the rectangle (or lines you just added) and click the Make Invisible button.
14. Click the Visibility Mode button to see the invisible objects (they will be transparent).
15. Create a new state and call it Closed.
16. Select the arc and the vertical lines and click the Make Invisible button. Select the rectangle (or lines) you just added and click the Make Visible button.
17. Test the block.
18. Save the block, and test it outside Block Editor.
19. Save and close the file.

PRACTICE 20-3

Dynamic Blocks – Wide Flange Beams

1. Start AutoCAD 2014.
2. Open **Practice 20-3.dwg**.
3. Create a block and name it Wide Flange Beam. Select the lower-left corner to be the base point, and make sure the block unit is Inch and the Open in block editor checkbox is on.

4. Add four linear parameters and name them as follows:

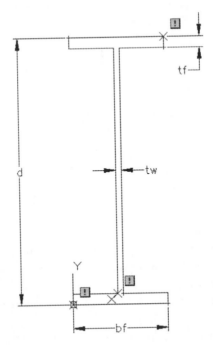

5. Select all linear parameters and set the grips to 0.
6. Add a scale action and select the d parameter and all objects.
7. Using the following table set the Value Set to List and input the following values (the first value is already set):

d (in)	tw (in)	bf (in)	tf (in)
44.02	1.02	15.95	1.77
34.21	1.63	15.865	2.95
27.63	0.61	10.01	1.10

8. Going to Parameter Sets tab, select lookup set, and add to the drawing, selecting the proper location for it.
9. Right-click the lookup action icon and select the Display Lookup Table option. A dialog box will open. Click the Add Properties button, and

you will see a list of all parameters added previously. Select them all and click OK.

10. Using the table above select all the matched values together, and name the first W 44 × 335, the second W 30 × 477, and the third W 27 × 129.
11. Test the block.
12. Save, and test the block outside Block Editor.
13. Save and close the file.

20.9 USING CONSTRAINTS FOR DYNAMIC BLOCKS

- In Block Editor, you can replace parameters and actions using geometric and dimensional constraints. This will make your life easier as it will simplify the steps and reduce the time it takes to produce dynamic blocks. Using geometric and dimensional constraints for creating dynamic blocks is the same as using them for ordinary objects, although there are some parameters such as visibility and some actions such as array that can't be replaced using geometric and dimensional constraints.
- One of the things you can do with dimensional constraints is set the value set such as the linear parameters using an increment or list.
- The interface is similar to what we learned in Chapter 14, with only three new buttons:
  - In the Dimensional panel in Block Editor there is a button called Block Table.
  - In the Manage panel in Block Editor there is a button called Construction.
  - In the Manage panel as well in Block Editor there is a button called Constraint Status.

20.9.1 Block Table Button

- This button is a replica of the lookup parameter and action, and it works with dimensional constraints in the same way. So why was this feature added? The answer is simple: you can't use the lookup parameter and action on dimensional constraints, so we need this feature to produce a popup list to select sizes, dimensions, etc. To use this feature, while you are in Block Editor, locate the **Dimensional** panel and click the **Block Table** button:

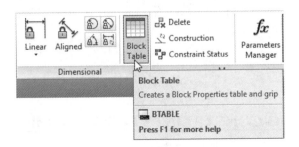

- The following prompts will appear:

```
Specify parameter location or [Palette]:
Enter number of grips [0/1] <1>:
```

- You will be asked to specify a location for the list to appear, then to specify the number of grips to be included.
- The following dialog box will appear:

- To add dimensional constraints to the list, click the following button:

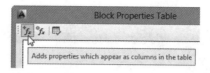

- The following dialog box will appear:

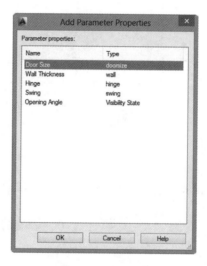

- Select the desired dimensional constraints, then click OK. You will see the following:

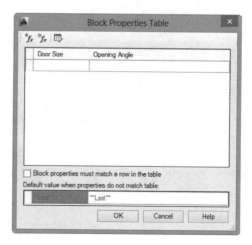

- Choose the first field in the first row and select one of the available values:

- Keep inputting values until you get something like the following:

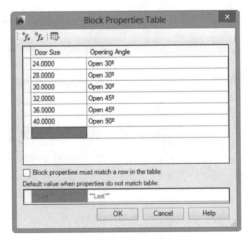

- In order for these values to appear here, you should go to the Properties palette of each dimensional constraint and add a list or increment.

- The other two buttons at the top left of the dialog box are used to create user parameters. You will see the following dialog box:

- The third button is used to correct the block table, if needed. If everything is fine, you will see the following message:

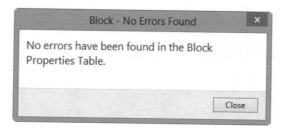

- But if AutoCAD finds any mistake (like one of the fields is empty) you will see something like the following message:

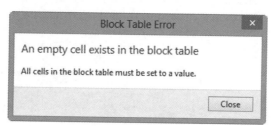

20.9.2 Construction Button

- A construction object is an object that you used to define your dynamic block. Inside Block Editor, locate the **Manage** panel, and click the **Construction** button:

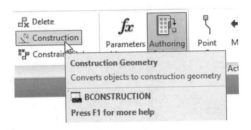

- You will see the following prompt:

```
Select objects or [Show all/Hide all]:
```

- Select the desired object(s).

20.9.3 Constraint Status Button

- This button allows AutoCAD to see your current geometric constraint status. There two options: either blue, which means partial constraints, or magenta, which means full constraints. This is an on/off button. It is best to keep it switched on to monitor your blocks. If you add all of your constraints and you still get blue, you probably need to add a fix constraint to one of the points, which is an AutoCAD requirement. To switch this button on/off, while you are inside the Block Editor, locate the **Manage** panel, and then select the **Constraint Status** button:

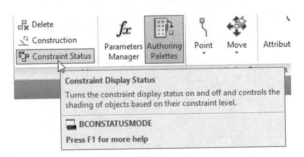

PRACTICE 20-4

Dynamic Blocks Using Constraints – Wide Flange Beams

1. Start AutoCAD 2014.
2. Open **Practice 20-4.dwg**.
3. Redo Practice 20-3 using geometric and dimensional constraints instead of parameters and actions.
4. Save and close the file.

PRACTICE 20-5

Dynamic Blocks Using Constraints – Creating a Window

1. Start AutoCAD 2014.
2. Open **Practice 20-5.dwg**.
3. Create a new block from the existing shape and name it **Window Elevation**, using the lower-left corner as the base point and Units to be Inch, making sure that Open in block editor is on.
4. Locate the Geometric panel, select the AutoConstrain button, and select all objects to define the geometric constraints, then hide all of them.
5. Using dimensional constraints set the following linear constraints, keeping the names used below:

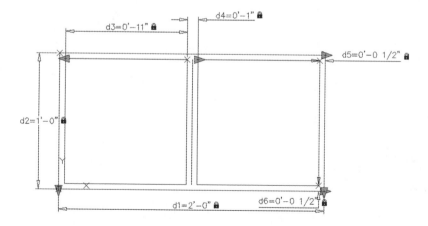

6. Using the **Parameters Manager**, make the following changes to the names: d1=Width, d2=Height, and d3=Glazing.
7. Set the following equations:
   a. Width = Height * 2
   b. Glazing = (Width/2)-d4
8. Select all dimensional constraints except Height, and set Grips = 0.
9. Set the value set for Height to be Increment, set the min = 1', max = 2', and increment = 6".
10. Set the line between the two glazing to be Construction object.
11. Test the block.
12. Save the block, and test it outside Block Editor.
13. Save and close the file.

NOTES:

CHAPTER REVIEW

1. You can add an action to a dimensional constraint.
   a. True
   b. False
2. The name of the palette in the Block Editor is _____.
3. Which one of the following parameters doesn't need an action?
   a. Linear
   b. Visibility
   c. Lookup
   d. Rotation
4. The rotate action works only with the rotation parameter.
   a. True
   b. False
5. While you are at the Properties palette of a linear parameter, you can set the value set to:
   a. None, Increment
   b. None, List
   c. Increment, List
   d. None, Increment, and List
6. All actions need parameters.
   a. True
   b. False
7. You can test a block inside the Block Editor before you save any changes.
   a. True
   b. False

CHAPTER REVIEW ANSWERS

1. b
3. b
5. d
7. a

Chapter 21 BLOCK ATTRIBUTES

In This Chapter
◇ What are block attributes?
◇ How to create a block attribute
◇ Editing block attributes and values
◇ How to extract attribute values from drawings

21.1 WHAT ARE BLOCK ATTRIBUTES?

- In the previous chapter, we discussed how to add dynamic features to blocks for more versatility, but dynamic blocks still don't have any data. Using block attributes we can include information inside a block:
  - If you create the block that contains attributes, you will define the attribute definitions, including Color, Cost, Height, and Width.
  - If you are inserting the block, you will input values (e.g., Red, 12.99, 3'-6", and 2'-6").
- The steps for adding block attributes are very simple:
  - Draw the graphical shape of the block.
  - Using the Attribute Definition command, define the attributes you want to collect data for.
  - Using the Block command (the normal command used to create blocks,) define the block that contains the attributes.
  - Insert the block and fill in the values.
  - Set the visibility of the attributes.
  - Edit the attributes either as values or definitions.
  - Extract the data to a table inside the drawing, or as an Excel or database file.

21.2 HOW TO DEFINE ATTRIBUTES

- This is the first step in creating the attribute definitions. To issue this command, go to the **Insert** tab, locate the **Block Definition** panel, and then select the **Define Attributes** button:

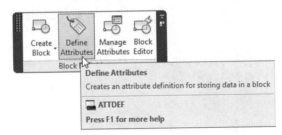

- The following dialog box will appear:

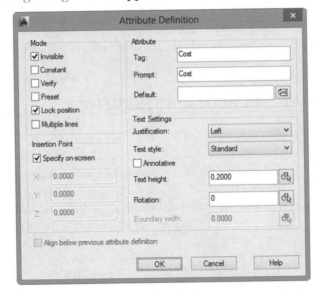

- There are four sections in this dialog box, discussed as follows.

21.2.1 Attribute Section

- This section contains three different settings:
  - Tag: the attribute name; you can select any desired name.
  - Prompt: the message that appears asking you for a piece of information.
  - Default: if there is a value that will be repeated more than others, this is the place to input it. If this value is connected to a field, then click the button at the right and input his field.

21.2.2 Mode Section

- There are six modes for the attribute definitions:
  - Invisible: allows you to create an invisible attribute that doesn't appear in the drawing.
  - Constant: allows you to create a constant value for all of the insertions.
  - Preset: allows you to create an attribute that will be equalized to the default value.
  - Verify: you will be asked to input the attribute twice to verify the value (applies only for Command window input).
  - Lock position: this mode will lock the position of the attribute definition.
  - Multiple lines: allows you to input multi-line text rather than single line text.

21.2.3 Text Settings Section

- This section controls the appearance of the attribute definition:
  - Justification: allows you to specify the justification of the text.
  - Text style: specifies the text used to input the attribute definition (it is best to create a specific text style prior to this step).
  - Whether the text will be annotative or not.
  - Text height: the text height to be used if the text style height is 0.
  - Rotation: the rotation angle of the attribute definition.
  - Boundary width: used only when multiple lines is selected. It will be the line length for the multiple lines.

21.2.4 Insertion Point Section

- This section is used to specify the position of the attribute definition either by typing the coordinates or by selecting the **Specify on-screen** option. You will have the choice of aligning the new attribute with the previously inserted attribute. Make sure the following checkbox is on:

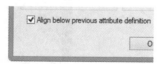

- After you define the attributes, start the Block command to create a block from the graphics and the attributes. While selecting the attributes, select them in the same order you would like to see them.

21.3 INSERTING BLOCKS WITH ATTRIBUTES

- You should control how the attribute questions will appear to you, and system variable ATTDIA will do the job. There are two values to choose from:
  - If **ATTDIA = 0**, questions will be shown at the Command window.
  - If **ATTDIA = 1**, questions will appear in a dialog box, just like the following:

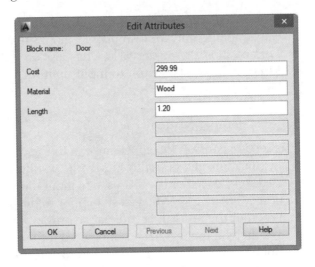

21.4 HOW TO CONTROL ATTRIBUTE VISIBILITY

- When you defined the attributes, you specified that some of the attributes will be visible and some will be invisible. This section discusses how to change this status temporarily. In order to control attribute visibility go to the **Insert** tab, locate the **Block** panel, extend the panel, and click the list:

- As displayed above, you have three choices to choose from:
  - **Retain Attribute Display**, which means AutoCAD will show the visible attributes and hide the invisible ones.
  - **Display All Attributes**, which means AutoCAD will show all attributes.
  - **Hide All Attributes**, which means AutoCAD will hide all attributes.

PRACTICE 21-1

 Defining and Inserting Blocks with Attributes

1. Start AutoCAD 2014.
2. Open **Practice 21-1.dwg**.
3. Zoom to the shape at the left.
4. Define the following attributes, inserting them below the door shape:

Tag	Material	Cost	Color
Prompt	Material	Cost	Color
Default	Wood	299.99	Dark
Invisible	ü	ü	ü
Preset	û	ü	û
Justification	Left	Align	
Text Style	Arial10	Align	

5. Define a new block and name it "Door," selecting the lower-left corner of the door as the insertion point, selecting all objects except attributes, and then selecting attributes one by one from the top to bottom. Click OK to finish.
6. Make layer A-Door current.
7. Using the ATTDIA command, make sure the value is 1.
8. Insert the new block into the three openings available in the architectural plan using the following information:
   a. For the one at the left, leave the default values.
   b. For the one in the middle, change the material to Aluminum.
   c. For the one at the right, change the cost to 399.99 and the color to Light.

9. Change the visibility mode to show all attributes.
10. Save and close the file.

21.5 HOW TO EDIT INDIVIDUAL ATTRIBUTE VALUES

- This section discusses how to edit individual block attributes. You can edit not only the values, but many other things as well. There are two ways to reach this command:
  - Double-click an existing block with attributes.
  - Go to the **Insert** tab, locate the **Block** panel, click **Edit Attribute**, and then select the **Single** button:

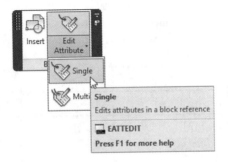

- You will see something like the following:

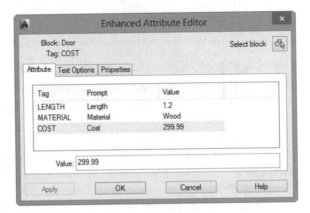

- There are three tabs: Attribute, Text Options, and Properties. You will see the Attribute tab when you start editing. In this tab, you can change the value of the attribute for this specific block. Click the **Text Options** tab to see the following:

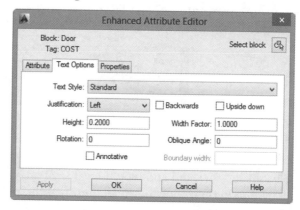

- As you can see you can change everything related to the text style for this specific block and attribute. Select the **Properties** tab to see the following:

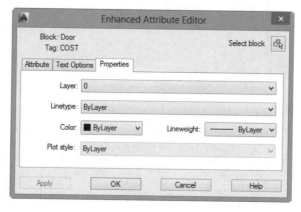

- You can change anything related to properties such as layer, linetype, color, and lineweight.

21.6 HOW TO EDIT ATTRIBUTE VALUES GLOBALLY

- Above we discussed how to edit individual block attributes. But what if we have many incidences of a block and want to change one specific value of the block? Using the above method the process will be lengthy and tedious. AutoCAD offers us a very simple command, the familiar Find/Replace command found in most software. To issue this command, go to the **Annotate** tab and locate the **Text** panel. You will find a field with ***Find text*** inside it:

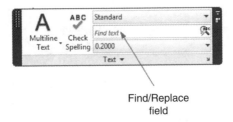

Find/Replace
field

- Click inside the field and type the value of attribute you want to change globally, and then click the small button at the right. You will see the following dialog box:

- Extend the dialog box by clicking the small arrow at the lower-left corner of the dialog box. You will see something like the following:

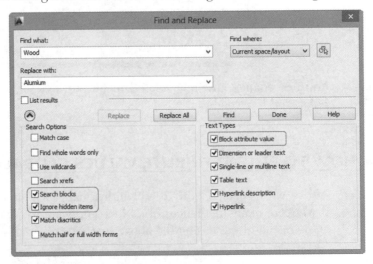

- In the above extension, make sure of the following:
  - Under Text Types, make sure the Block attribute value is on.
  - Under Search Options, make sure that Search blocks is on.
  - Under Search Options, make sure that Ignore hidden items is off.
- Finally, click one of the three execution commands: Find, Replace, Replace All.

PRACTICE 21-2

 ### Editing Attribute Values Individually and Globally

1. Start AutoCAD 2014.
2. Open **Practice 21-2.dwg**.
3. There have been some mistakes in inputting the attribute values, and we need to unify them. Change the following:
   a. All doors are wood, and none is aluminum.
   b. All doors are 299.99, and none is 399.99.
   c. All doors are dark, and none is light.
4. Double-click any two doors and make sure the changes were made.
5. Show all attributes.
6. Double-click the main door, then select the Cost attribute.
7. Go to Text tab and set the Text style to be Standard and set the Height = 0.15.
8. Go to the Properties tab and select Text layer.
9. Click OK, and compare this attribute to the others.
10. Retain Attribute Display.
11. Save and close the file.

21.7 HOW TO REDEFINE AND SYNC ATTRIBUTE DEFINITIONS

- We have already discussed how to edit values of attributes. But what if there is a missing attribute you need to add or one you need to remove? What if you want to edit the modes of an existing attribute? To do these things, you will need to use the Block Attribute Manager. To issue this command, go to the **Insert** tab, locate the **Block Definition** panel, and then select the **Manager Attributes** button:

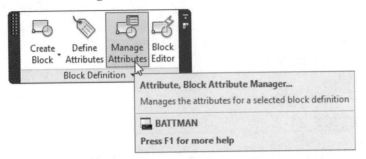

- You will see the following dialog box:

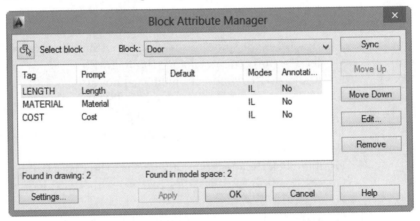

- There are two ways to start working:
  - Either click the button at the top right to select a block
  - Or click the list to the right of the button to select the desired block
- You will see a list of attributes for the selected block. You can do four things with these attributes:
  - Change the position of the attribute in relation to other attributes by moving the selected attribute one position up.
  - Change the position of the attribute in relation to other attributes, by moving the selected attribute one position down.
  - Remove (delete) the selected attribute.

- Edit the definition of the selected attribute. You will see the following dialog box:

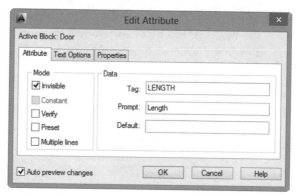

- In this dialog box, you will be able to edit Tag, Prompt, Default value, Mode, and all other text options and properties as well.
- What about adding a new attribute to an existing block? Take the following steps:
  - Insert the desired block in your drawing.
  - Explode it.
  - The existing attributes will appear.
  - Define the new attribute.
  - Redefine the block using the same name.
- When we edit, delete, or add new attributes, it will affect the new insertions of the block, but what about the blocks already inserted? We need to synchronize in order to see the effects of the changes. To issue this command, go to the **Insert** tab, locate the **Block Definition** panel, and then select the **Synchronize** button:

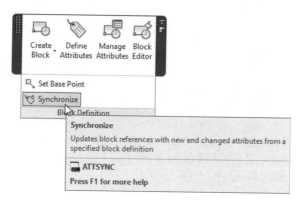

- You will see the following prompt:

  ```
  Enter an option [?/Name/Select] <Select>:
  ```

- Type the name of the block, or choose the Select option to select it, and you will see the following prompt:

  ```
  ATTSYNC block Window? [Yes/No] <Yes>:
  ```

- If you input Yes (default), AutoCAD will start the synchronize process. The newly added attributes will be empty, but you can fill them by double-clicking the desired block.

PRACTICE 21-3

Redefining Attribute Definitions

1. Start AutoCAD 2014.
2. Open **Practice 21-3.dwg**.
3. Using the Block Attribute Manager, do the following:
   a. Remove the Color attribute
   b. Make the Cost attribute modes only Invisible
   c. Change the cost to appear first in the list of questions
4. Insert block Door in an empty space in the drawing, then explode it.
5. Make layer 0 current.
6. Add a new attribute with the following data:
   a. Tag: SUP
   b. Prompt: Supplier
   c. Default: MYNE Wood
   d. Modes: Invisible
   e. Insert it below the existing two attributes
7. Redefine the block under the same name using the same base point (very important).
8. Double-click any of the blocks, and you will see that the Supplier attribute was not added because the sync command was not issued.
9. Issue the Synchronize command, and select one of the doors.
10. Now double-click one of the blocks again. Do you see the Supplier attribute? _____
11. Save and close the file.

21.8 HOW TO EXTRACT ATTRIBUTES FROM FILE(S)

- After doing the above, now we need to know how to extract the data and put it either in a table inside the same drawing, or another drawing, or create an Excel worksheet or Access database file. Extracting data used to be a very complicated procedure that involved creating external files as templates, but not anymore. Extracting data in AutoCAD is now done through a wizard that walks you through the process. To access this command, go to the **Insert** tab, locate the **Linking & Extraction** panel, and then select the **Extract Data** button:

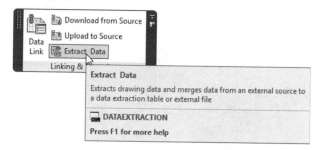

- The following dialog box will appear:

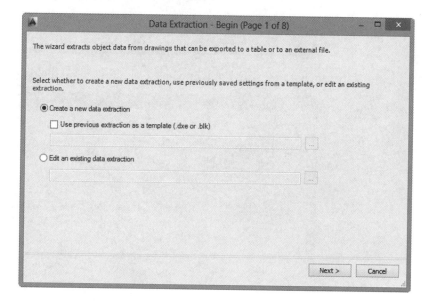

- You can choose from one of the three available options:
  - Create a new data extraction
  - Use the previous extraction as a template
  - Edit an existing data extraction
- Let's assume we want to create a new extraction. We will select the first option, and click the Next button. AutoCAD will ask you to save this extraction (file format is *.dxe) for future use, and the following dialog box will appear:

- After saving the .dxe file, you will see the following dialog box:

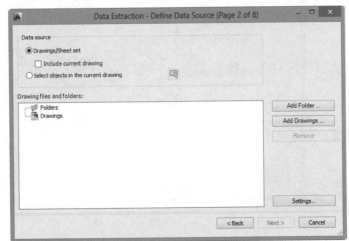

- AutoCAD needs to know where to find the files to extract the data from. You have two choices to choose from:
  - Certain files (or sheet sets) to include or to exclude the current file
  - Select objects in the current drawing
- Selecting the first choice means you have to specify the folders and drawing files using the two buttons at the right. Clicking the **Add Folder** button, you will see the following dialog box:

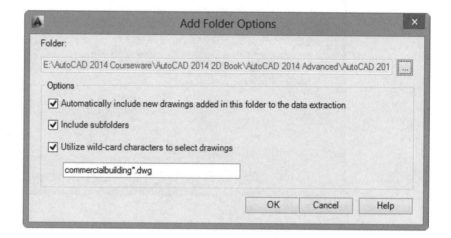

- There are four things to control in this dialog box:
  - Click the small button with three dots inside to specify the folder that contains the desired files.
  - Select whether or not you want to "Automatically include new drawings added in this folder to the data extraction."
  - Select whether or not you want to include all subfolders of the selected folder.
  - Select whether or not you want to "Utilize wild-card characters to select drawings." This is similar to what we can do in the Windows Search facility. For instance, Commercial Building*. dwg means you want to select all files starting with Commercial Building.

- If you want to go with the second choice, simply click at the right to select the desired objects you want to extract data from. If you create a wild card, you may see something like the following:

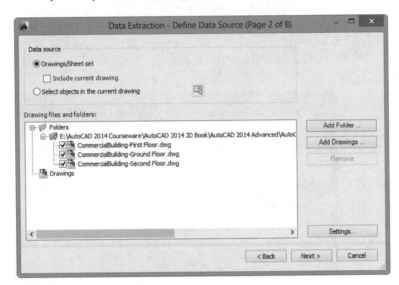

- When done click the Next button, and you will see the following dialog box:

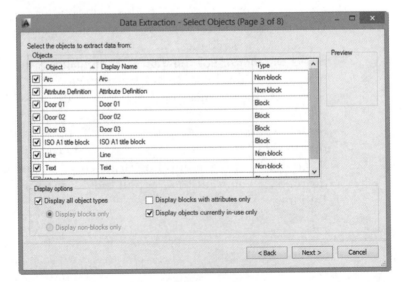

- As shown above, the list shows objects and blocks. In order to refine this list, use the checkboxes beneath. You have three choices:
  - Select whether or not to Display all object types. If you turn this off, you have to select whether to Display blocks only or to Display non-blocks only.
  - Select whether or not to Display blocks with attributes only.
  - Select whether or not to Display objects currently in-use only.
- The example below shows only blocks with attributes, and we deselected one of the blocks that we don't want to include in the extraction process:

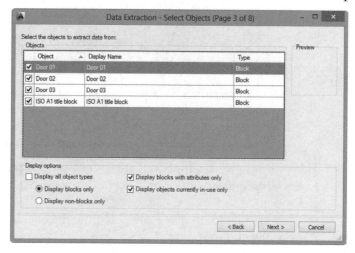

- When done, click the Next button, and you will see the following dialog box:

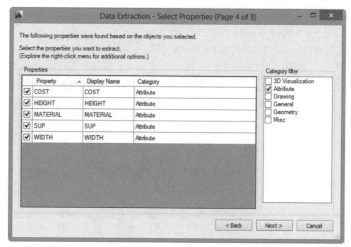

- At the right and under the Category filter, deselect all choices except Attribute to filter the information needed. You can also further deselect the attributes you don't want to include in the extraction process. When done click Next to continue to the fifth step in the extraction wizard. You will see the following dialog box:

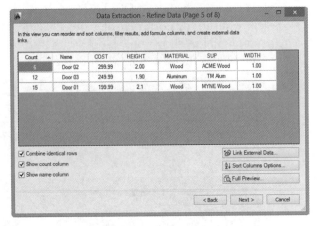

- There are some basic functions you can do on this page:
  - Sort the table ascending or descending by clicking the desired column header.
  - Change the location of any column by clicking the header and dragging and dropping.
  - Change the column width by dragging the line separating any two columns.
  - You can select to Combine identical rows.
  - Show count column or not.
  - Show name column or not.
- If you right-click any header, you will see the following menu:

- Use this menu to do all or any of the following:
    - Rename a column.
    - Hide and Show columns.
    - Set Column Data Format. You will see the following dialog box:

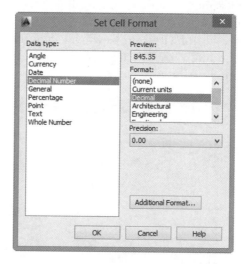

- Insert a new formula column. You will see the following dialog box (in this dialog box type the name of the new column, then choose the names of the attributes you want to use by using drag-and-drop, using the arithmetic function between them):

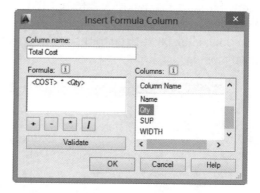

- Edit and Remove Formula Column.
- Insert Totals Footer.
- Create a filter for the selected column.

- If you want to establish a link between this table and an Excel file, click the Link External Data button at the right.
- If you want to set advanced sorting criteria, click the Sort Columns Options button at the right.
- If you want to establish a data link between these extracted data and Excel sheet.
- The last button (Full Preview) will show a preview of the table.
- When done, click Next to go to the next step, and you will see the following dialog box:

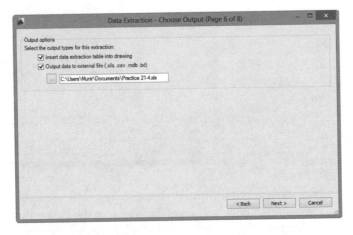

- On in this page can select to insert the data extraction as a table in the current file and produce an external file (e.g., an Excel worksheet or Access database file).
- When done click Next to go to the seventh step, and the following dialog box will appear:

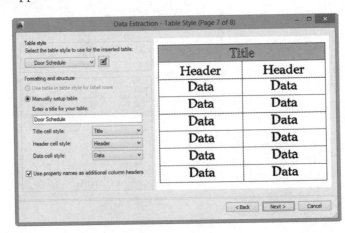

- Select the table style you want to use to insert the data extraction in the current file. When done click Next to go to the last step, and the following dialog box will appear:

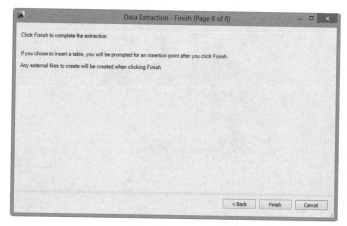

- Click Finish to wrap up the extraction process.

PRACTICE 21-4

Extracting Attributes

1. Start AutoCAD 2014.
2. Open **Practice 21-4.dwg**.
3. Extract data from the following three files, which can be found in your Chapter 3 practices folder:
   a. CommercialBuilding-Ground Floor.dwg
   b. CommercialBuilding-First Floor.dwg
   c. CommercialBuilding-Second Floor.dwg
4. This is the final product of the extraction. Do whatever required to produce an identical replica:

Door Schedule							
Name	HEIGHT	WIDTH	MATERIAL	Supplier	Qty	COST	Total Cost
Door 01	2.1	1.00	Wood	MYNE Wood	15	199.99	$2,999.85
Door 02	2.00	1.00	Wood	ACME Wood	6	299.99	$1,799.94
Door 03	1.90	1.00	Aluminum	TM Alum	12	249.99	$2,999.88
					33		$7,799.67

5. Save and close the file.

NOTES:

CHAPTER REVIEW

1. While you are defining a new attribute you can make the default value equal to a field.
   a. True
   b. False
2. Which one of the following statements is *not* true about the extracting process?
   a. While you are extracting you can add a formula column.
   b. You can add more than one formula column.
   c. The Formula column will use attributes and fields.
   d. You can add a total footer.
3. You can edit attribute values by double-clicking the desired block.
   a. True
   b. False
4. After inserting a block containing attributes in a drawing you can do all but which of the following?
   a. Edit the values one by one
   b. Edit the attribute definitions
   c. Allow AutoCAD to synchronize the values automatically
   d. Add a new attribute to the block
5. _____ mode allows you to create an attribute that will be equalized to the default value.
6. Invisible and constant are _____.
7. _____ is a system variable that will control whether to show or not show a dialog box while inputting values for attributes.

CHAPTER REVIEW ANSWERS

1. a
3. a
5. Preset
7. ATTDIA

22 EXTERNAL REFERENCING (XREF)

Chapter

In This Chapter

◇ What is an External Reference (Xref)?
◇ Inserting Xrefs
◇ Xref Layers
◇ Editing Xrefs
◇ Xref Functions
◇ Xref Clipping
◇ Xref Special functions
◇ eTransmit command

22.1 INTRODUCTION TO EXTERNAL REFERENCE

- Today's working world is collaborative, and we need to be able to share work. External references (or Xrefs) help engineers/draftsmen collaborate and coordinate their drawings by allowing users to import other drawings for use while segregating the content of each drawing to keep it separate and organized. This is not a new feature of AutoCAD, but the last two or three versions of AutoCAD only allowed importing file formats such as DWF, DGN, PDF, and image formats.
- Xref files are linked to the original files so when there is a change in the original file, AutoCAD notifies you. When you open an Xref file, the latest version of the file is loaded into your drawing.
- Using Xref files is much more efficient than importing DWG files, as this method will add the whole file into your drawing, you will not be able to update it and the layers and blocks will be mixed in with your current layers and blocks.

22.2 INSERTING EXTERNAL REFERENCE FILE FORMATS

- There is one command that can insert all five types of files in the current drawing file. To reach this command, go to the **Insert** tab, locate the **References** panel, and then select the small arrow at the lower right:

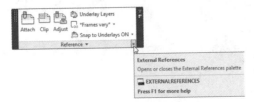

- The **External Reference** palette will appear showing only the name of the current file:

- To attach any file format, simply go to the top-left of the palette, where a small list button exists, and click it. You will see something like the following:

- As shown above, you can select one of the following:
  - Attach DWG file
  - Attach Image file

- Attach DWF file
- Attach DGN file (Microstation V8 file)
- Attach PDF file
- Attach Point Cloud

22.2.1 Attach a DWG File

- You will use this command to attach an AutoCAD drawing in the current drawing. Once you click OK, you will see the following dialog box:

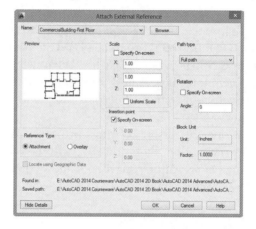

- As shown above four parts of this dialog box are identical to the Insert Block command:
  - Scale
  - Insertion point
  - Rotation
  - Block unit
- Path type and Reference type are new.
- For Path type, we already said that AutoCAD keeps a link to the original file to keep track of any changes to the file. In order to do that, AutoCAD saves the current path of the attached DWG file. AutoCAD will ask you what type of path to use:

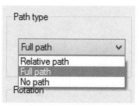

- **Full path**: AutoCAD will save the exact path of the attached DWG file, remembering the drive and folder(s), which means if the file is moved to a different place, AutoCAD will report the error.
- **Relative path**: The default option. AutoCAD will only save the name of the folder containing the file, ignoring the other folders and drive. So if your folder is moved, AutoCAD will be able to locate it.
- **No path**: AutoCAD will not save the path for the DWG file, but as an alternative, it will search the current folder, then the search paths discussed in the Options dialog box, the Files tab, and both paths discussed under Project and Support.
- The Reference Type is for people using Xrefs to import DWG files. Let's assume that person B brings in a DWG to his/her file from person A. Then person C brings in person B's file. What will person C see in his/her file?
  - Attachment: If person B brings in person A's drawing and chooses the Attachment option, then person C will see both A's drawing and B's drawing.
  - Overlay: If person B brings in person A's drawing and chooses the Overlay option, then person C will see only person B's drawing.
- As a final note, AutoCAD will not allow you to reference the same DWG file twice in the same drawing.

22.2.2 Attach an Image File

- You will use this command to attach an image file in the current drawing. You will see a normal Open dialog box to select your desired file, and once you click OK, you will see the following dialog box:

- We already discussed all parts of this dialog box. AutoCAD allows you to reference the same file twice in the same drawing.

22.2.3 Attach a DWF File

- You will use this command to attach a DWF file in the current drawing. You will see a normal Open dialog box to select your desired file, and once you click OK, you will see the following dialog box:

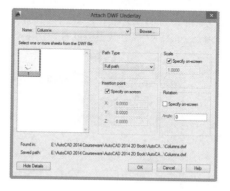

- This dialog box is identical to the attaching DWG file dialog box except the part where you select which page of the DWF file you want to attach, that is, if the file has multiple pages. AutoCAD allows you to reference the same file twice in the same drawing.

22.2.4 Attach a DGN File

- You will use this command to attach a DGN (Microstation) file in the current drawing. You will see a normal Open dialog box to select your desired file, and once you click OK, you will see the following dialog box:

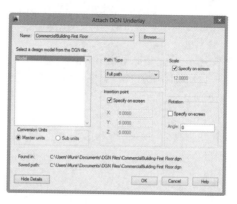

- This dialog box is identical to the one for DWG files except for one part: Conversion Units. If you know that Microstation files deal with two units, Master and Sub, you have to tell AutoCAD the current AutoCAD unit (Master or Sub). AutoCAD allows you to reference the same file twice in the same drawing.

22.2.5 Attach a PDF File

- You will use this command to attach a PDF file in the current drawing. You will see a normal Open dialog box to select your desired file, and once you click OK, you will see the following dialog box:

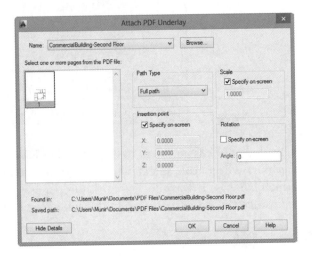

- This dialog box is identical to the attaching DWG file dialog box except the part where you select which page of the PDF file you want to attach, that is, if the file has multiple pages. AutoCAD allows you to reference the same file twice in the same drawing.
- Note the following:
  - The reference file will always be located in the current layer, so be sure not to freeze this layer.

- The first time you insert a reference file of any type, you will notice at the right part of the status bar an icon called **Manage Xrefs**. Clicking it once will show the External Reference palette:

22.3 EXTERNAL REFERENCE PALETTE

- Mastering the different parts of this palette will be very helpful to you. We will discuss each part of the palette and what it does. The upper part has two views:
  - List view shows all the files attached to your file. Each attachment has a different icon to its left to distinguish it from the others. To the right of it, you will see information such as its status, size, path, etc.:

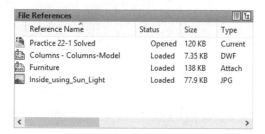

  - Tree view will show the same files but in tree view. This is very handy if your referenced DWG file has reference files by itself. This view has no details:

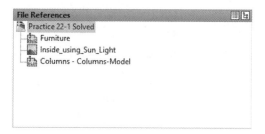

■ The lower part of the palette also has two views:

- Either Detail will show something like the following. You can see that the Saved Path is editable and can be changed:

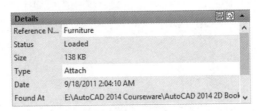

- Or, you will see a preview of the attached file, something like the following:

22.4 USING THE ATTACH COMMAND

■ AutoCAD offers another way to attach all types of files without using a palette: the Attach command. To access this command, go to the **Insert** tab, locate the **Reference** panel, and then select the **Attach** button:

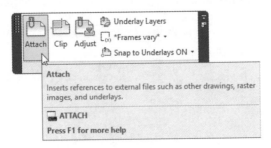

- The following dialog box will appear to first select the first type of file (use the list shown below) and then to select the desired file:

22.5 REFERENCE FILES AND LAYERS

- External Reference files, as discussed, are segregated from the original layers. To check this out, go to the Layer Properties Manager palette, and you will see something like the following:

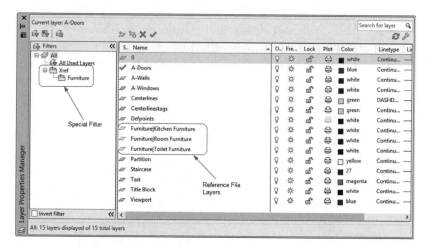

- In the above palette, you can easily distinguish between the original layers of the current file and the layers of the DWG reference file. At the left, you can see an automatic filter called Xref.
- Layers from a DWG reference file always start with the name of the file, then a pipe (|), and then the name of the layer: *filename|layername*.
- There are two important facts about layers in a reference file:
  - You cannot draw on reference layers, and you can't make any of them current.
  - AutoCAD allows you to change the properties of referenced layers (color, linetype, lineweight, etc.) and to change the status of a layer (on/off, freeze/thaw, etc.) without affecting the original file. Once you save the file, the changes made and the status will be saved. If you want to change this setting, go to the Application menu, select **Options**, go to the **Open and Save** tab, and then click off the Retain changes to Xref layers checkbox, as shown below (you can do the same thing using the system variable VISRETAIN by setting its value to be 0 instead of 1):

22.6 CONTROLLING THE FADING OF A REFERENCE FILE

- When you attach a DWG file to your current file, you will notice it is faded. This is true because the default setting for any DWG reference file is 70% as it shown in the Reference panel.
- You can do two things with this setting:
  - Control the fading percentage by increasing it or decreasing it using the Reference panel:

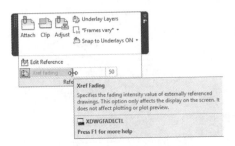

- Turning on/off the fading (if you don't want AutoCAD to add fading to your DWG reference file) using the Reference panel:

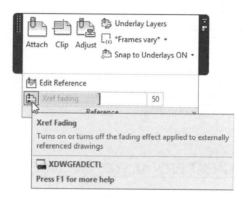

PRACTICE 22-1

Attaching and Controlling Reference Files

1. Start AutoCAD 2014.
2. Open **Practice 22-1.dwg**.
3. Make sure that layer 0 is the current layer.
4. UsingXref, attach the Furniture.dwg file using 0,0.
5. UsingXref, attach Columns.dwf using 0,0.
6. Go to the D-Size Architectural Plan layout.
7. UsingXref, attach Inside_using_Sun_Light.jpg using a base point and scale, go to the frame, and insert it in the upper part of the legend using proper scale to fit within.
8. Close the External Reference palette.
9. Start the Layer Properties Manager.
10. How many layers were added with Furniture.dwg? _____
11. What are their names? _____, _____,
   _____
12. Change the color of Toilet Furniture to Red.
13. Freeze layer Kitchen Furniture.
14. Go to the Options dialog box, select the Open and save tab, and turn off the checkbox Retain changes to Xref layers.
15. Save the file and close it.

16. Reopen the file. What happened to the reference file layers? _____
    _____
    _____ Why? _____

17. Turn off the fading of the reference DWG file, then turn it on to see the difference.

18. Click the Manage Xrefs button at the status bar to show the External Reference palette, then examine the four parts of the palette.

19. Save and close the file.

22.7 EDITING EXTERNAL REFERENCE DWG FILES

- AutoCAD allows you to edit any Xref DWG file using two methods:
  - Using the Edit Reference command
  - Using the Open command
- You should not edit another person's file without his/her permission.

22.7.1 Using the Edit Reference Command

- This command allows you to edit the Xref DWG file in its place. To issue this command, go to the **Insert** tab, locate the **References** panel, and then select the **Edit Reference** button:

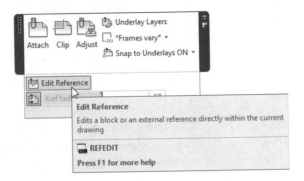

- You will see the following prompt:

```
Select reference:
```

- Click on the desired DWG to edit it, and you will see the following dialog box:

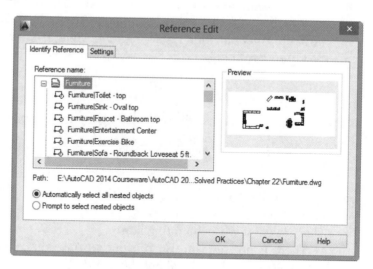

- As you can see the filename of the file selected appears with all of its blocks, which proves that this command can edit not only Xref files, but blocks as well. If this is the file you want to edit click OK. Once you do two things will take place:
  - All of the drawing will be dimmed except the selected file.
  - A new context tab titled Edit Reference will appear showing four buttons.
- Now you can use all the Modifying commands to alter the file in its place.
- The context tab will have four buttons:
  - Add to Working Set allows you to add more objects to the current set of objects to edit.
  - Remove from Working Set allows you to remove objects from the current set of objects to edit.
  - Save Changes: this is the final step after you finish editing
  - Discard Changes: this is the final step after you finish editing and decide not keep the changes.

22.7.2 Using the Open Command

- This command allows you to open the Xref DWG file for editing, then save the file with the new amendments. To do that, click any object in the Xref DWG file, then right-click, and a long menu will appear. Select the **Open Xref** option:

- The file will open for editing. Make the changes you need, then save the file and close it.
- There two things worth noting here:
  - There is no way you can prevent anyone who referenced your file from opening it and making changes, but you can prevent them from editing in place. To do that go to the Options dialog box, then select the Open and Save tab, and click off the Allow other users to Refedit current drawing option:

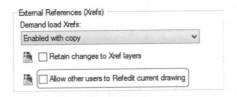

  - Regardless of the method of editing, if any changes were saved to the Xref file, you will see a balloon stating that changes took place to the file and you need to reload it, just like the following:

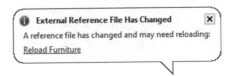

22.8 EXTERNAL REFERENCE FILE RELATED FUNCTIONS

- When you are working in the top part of the External Reference palette, you can right-click the name of the file to get the following menu:

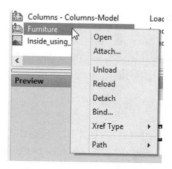

- As shown, there are six commands that you can issue for the selected file:
  - Open (already discussed)
  - Attach (will show you the Attachment dialog box to edit the settings of the attachment)
  - Unload
  - Reload
  - Detach
  - Bind
  - Xref Type (Attach or Overlay)
  - Path (Make Absolute, Make Relative, Remove Path)
- We will discuss these six commands in the following.

22.8.1 Unload Command

- Unloading an attached file means the file will disappear from the display, but the link inside the palette will stay. This is (in principle) similar to the freeze function in layers. The unloaded file will look like the following in the list:

Reference Name	Status	Size	Type	Date
Practice 22-1 Solved*	Opened	120 KB	Current	9/16/201
Columns - Columns-Model	Loaded	7.35 KB	DWF	9/16/201
Furniture	Unloa...		Attach	
Inside_using_Sun_Light	Loaded	77.9 KB	JPG	5/23/201

22.8.2 Reload Command

- The Reload command can be used to reload an unloaded file. Or, you can simply refresh all of your current links if needed.

22.8.3 Detach Command

- This command will make a file disappear just like the Unload command, except you will lose the link as well. So, you will not see the file in the list. If you need the file again, you have to attach it again.

22.8.4 Bind Command

- If for any reason, you opt to keep the latest copy of an Xref DWG file in the current file and cut the link between the two files, you should use the Bind command. The Bind command will bind the contents of the desired file inside your current drawing and will remove the link from the files list. The Bind command offers two different methods for binding. You will see the following dialog box:

- As shown the two choices are:
  - Bind
  - Insert
- The difference between the two methods is the naming of layers, styles, and blocks after the binding process ends. Assume we have a layer called:

Furniture|Frame

- If you use bind the name will become Furniture0Frame (the name of the layer still has the name of the file that it comes from and | was converted to 0).
- On the other hand, if we use Insert, it will become Frame.
- This applies to styles (text styles, dimension styles, etc.) and blocks.

22.8.5 Xref Type

- This option allows you to change the Xref type from Attach to Overlay and vice-versa.

22.8.6 Path

- Upon the attachment of Xref, you should specify if you want to make the path Absolute, Relative, or No Path. With this option, you can change this setting to any of the other options.

PRACTICE 22-2

External Reference Editing

1. Start AutoCAD 2014.
2. Open **Practice 22-2.dwg**.
3. Click on any object in Furniture.dwg, then right-click, and select the Open Xref option.
4. Erase the table and chairs in the kitchen and save the file.
5. A bubble will appear to reload the file. Reload the file to get the latest version of it.
6. Using the External Reference palette, unload Furniture.dwg.
7. Reload Furniture.dwg.
8. Open the Layer Properties Manager palette and take a look at the three layers from Furniture.dwg.
9. Using the External Reference palette, bind Furniture.dwg using the Bind option.
10. List the things that took place after binding:
    a. _____
    b. _____
    c. _____
11. Save and close the file.

22.9 CLIPPING AN EXTERNAL REFERENCE FILE

- AutoCAD allows you to clip an Xref file (of any type) in order to show some of it rather than all of it. To reach to this command, go to the **Insert** tab, locate the **Reference** panel, and then select the **Clip** button:

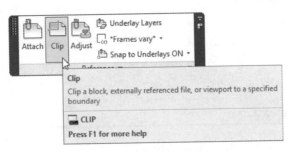

- The following prompt will appear:

```
Select Object to clip:
```

- You can select any type of Xref file. For simplicity we will show the DWG file prompts, since they are similar to the others. The prompts you will see are:

```
Enter clipping option [ON/OFF/Clipdepth/Delete/
generate Polyline/New boundary] <New>:
```

- The default option is to create a new boundary to clip the existing referenced file. So if you press [Enter] the second prompt will appear:

```
[Select polyline/Polygonal/Rectangular/Invert clip]
<Rectangular>:
```

- Using this prompt you can:
  - Select a polyline that you draw as a boundary
  - Define a new boundary using the Polygon option that allows you to draw any irregular shape
  - Define a new boundary using a rectangular shape
  - Invert an existing boundary
- Based on the option you select, you will define the desired boundary, and you will see some of the referenced file rather than all of it.

- The other options are:
  - Turn the clipping on/off while keeping it
  - Deleting the clipping boundary
  - Clip depth (which has to do with 3D)
- Note that by default you can't see the clipping boundary, due to the hidden nature of it. To show it, and accordingly manipulate it, go to the **Insert** tab, locate the **Reference** panel, check the second list at the right, and you will see:

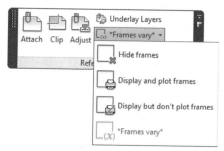

- There are four choices to choose from:
  - Hide frames
  - Display and plot frames
  - Display but don't plot frame
  - Frames vary
- While the frame is shown, you can click the frame itself to show something like the following:

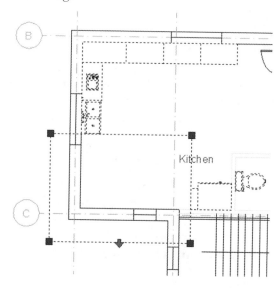

- With grips displayed you can:
  - Resize the frame
  - Invert the frame
  - Delete the frame (deleting the frame using the [Del] button will mean deleting the Xref file—be careful)

22.10 CLICKING AND RIGHT-CLICKING A REFERENCE FILE

- There are two actions you can take while you are working with an Xref file:
  - When you click it, a context tab will appear with related panels for editing, clipping, and other stuff.
  - When you click, then right-click, similar specific functions will appear with a menu relating to the type of referenced file selected.

22.10.1 Clicking and Right-Clicking a DWG File

- Clicking (one click) a DWG referenced file will open the External Reference context tab, which includes three panels, just like the following:

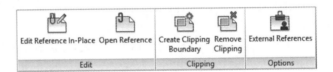

- The Edit panel contains Edit Reference In-Place and Open Reference buttons (both discussed) the Clipping panel contains the Create Clipping Boundary and Remove Clipping buttons (both discussed), and finally, the Options panel contains the External References button that will show External Reference palette.
- Right-clicking a DWG referenced file will show the following menu:

- These options are identical to the buttons in the context tab.

22.10.2 Clicking and Right-Clicking an Image File

- Clicking (one click) an image referenced file will open the Image context tab, which includes three panels, just like the following:

- The Adjust panel contains the Brightness, Contrast, and Fade slider to control the image colors, the Clipping panel contains the Create Clipping Boundary and Remove Clipping buttons, and finally, the Options panel contains the Show Image, Background Transparency (whether to show objects behind the image), and External References buttons.
- Right-clicking an image referenced file will show the following menu:

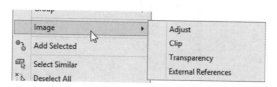

- These options are identical to the buttons in the context tab.

22.10.3 Clicking and Right-Clicking a DWF File

- Clicking (one click) a DWF referenced file will open the DWF Underlay context tab, which includes three panels, just like the following:

- The Adjust panel contains the Contrast and Fade sliders and the Display in Monochrome button to control the image colors, the Clipping panel contains the Create Clipping Boundary and Remove Clipping buttons, the Options panel contains the Show Underlay, Enable Snap, and External

References buttons, and finally, the DWF layers panel includes Edit Layers, which will show the following dialog box:

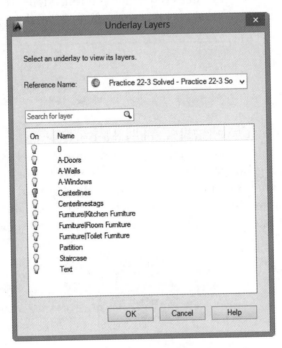

- Right-clicking a DWF referenced file will show the following menu:

- These options are identical to the buttons in the context tab.
- The DGN and PDF tabs are identical to those for DWF.
- Notice that there are two buttons in the Reference panel that we didn't touch on:
  - Underlay Layers button will show a dialog box of Edit Layers mentioned above for all DWF, DGN, and PDF attached to your current drawing.

- Snap to Underlays button allows you to acquire a point on DWF, DGN, and PDF files. You will see the following list:

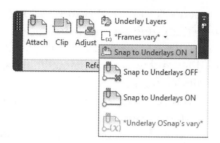

PRACTICE 22-3

 External Reference File Clipping and Controlling

1. Start AutoCAD 2014.
2. Open **Practice 22-3.dwg**.
3. Using the Insert tab and Reference panel, click Clip, click one of the pieces of furniture, select the rectangular shape, and create a rectangle around the kitchen and toilet.
4. What happened to the other furniture? _____
5. Select to show the frame of the clip.
6. Click the frame and shrink it to include only the kitchen.
7. Invert the clipping frame.
8. Using the Clip command delete the clipping frame.
9. Go to the D-Size Architectural Plan layout and zoom to the image at the top right of the title block.
10. Click the frame of the image. What happened? _____
    _____
11. Change the Fade % to 30%.
12. Using the context tab, clip the image to a smaller size.
13. Go back to Model space, select one of the columns, then right-click, and set DWF snap to off.
14. Hide the frames of the clipping.
15. Save and close the file.

22.11 USING THE eTRANSMIT COMMAND WITH EXTERNAL REFERENCE FILES

- If you attached several DWG files to your current file, along with images and DWF files, and you need to send to the file to someone, you will need to use the eTransmit command because your file is now dependent on other files and uses font, hatch, linetype, and shape files that may not exist on the recipient's computer. eTransmit allows you to collect all files related to your current file, zip them, and prepare them to be sent using e-mail or other means.
- Using the Application menu, select **Publish/eTransmit**. If you didn't save your file, you will see the following message:

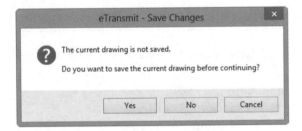

- Save your file by clicking Yes, then the following dialog box will appear:

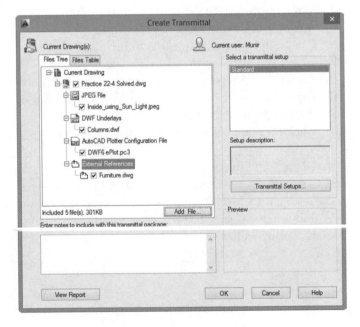

- At the top left, you can see two tabs. The default is Files Tree, and the other view is Files Table, like the following:

- You can include/exclude any undesired file. Using the button labeled **Add File** you can add other files to the existing list.
- Use the Transmittal Setups button to create your own parameters. Click the Transmittal Setups button, then click the New button, and you will see the following dialog box:

- Type the name of the new setup, then click the Continue button to see the following dialog box:

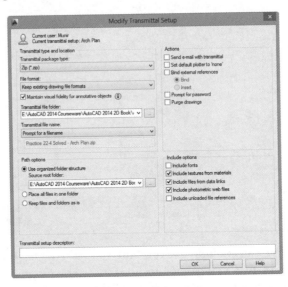

- All the options in this dialog box are very simple and should be easy to understand.
- Make the necessary changes, then click OK to end this dialog box.
- As a final step, click the button at the lower left labeled View Report to check everything, and you will see something like the following:

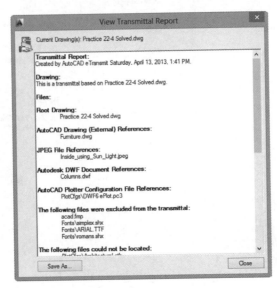

PRACTICE 22-4

 Using eTransmit

1. Start AutoCAD 2014.
2. Open **Practice 22-4.dwg**.
3. Start the eTransmit command.
4. See the Files Tree and Files Table. Is the AutoCAD Plotter Configuration File listed? _____ If yes, what is it? _____
5. Create a new Transmittal Setup and name it Architectural Drawing. Make sure of the following:
   a. Package type = Zip file
   b. File format = AutoCAD 2007
   c. Include fonts = yes
6. As you noticed, the fonts files were added.
7. View the report.
8. Save the file as Architectural Drawing.zip.
9. Save and close the file.

NOTES:

CHAPTER REVIEW

1. You can attach DOC and XLS files to an AutoCAD drawing using the External Reference palette.
    a. True
    b. False
2. There is only one type of reference type called "Overlay."
    a. True
    b. False
3. After you attach a DWF file, which one of the following is incorrect?
    a. You can control DWF layers
    b. You can Unload and Reload
    c. You can Bind
    d. You can Detach
4. You can attach the same image file more than once in the current file.
    a. True
    b. False
5. Which one of the following statements is *not* true?
    a. Once you attach one file an icon will appear at the right part of the status bar.
    b. The current layer will always be the location of the reference file.
    c. Layer 0 will always be the location of the reference file.
    d. Relative path is better than Full path.
6. If we have a layer in a reference DWG file named Furniture|Chair, after using the Bind command it will be _____.
7. _____ allows you to collect all related files in your current file, zip them, and prepare them to be sent using e-mail or other means.

CHAPTER REVIEW ANSWERS

1. b
3. c
5. c
7. eTransmit

Chapter 23 SHEET SETS

In This Chapter

◇ What are sheet sets?
◇ Types of sheet sets
◇ Understanding Sheet Set Manager
◇ Steps for using an example sheet set
◇ Steps for using existing drawings to create a sheet set
◇ eTransmit, Archive, and Publish sheet sets

23.1 INTRODUCTION TO SHEET SETS

■ The ultimate goal of any architectural or engineering company is to print out the final design in a sheet set, but printing individual drawings, gathering them into a set, and labeling them can be time consuming. International groups such as the U.S. National CAD Standard (NCS, http://www.nationalcadstandard.org) have standards that address sheet sets and specify discipline designator, sheet type, and sheet sequence numbers. Companies adopting NCS standards expect AutoCAD to adhere to these standards by providing sheet set creation and manipulation, and AutoCAD has provided this feature since 2005. AutoCAD allows you to create a sheet set and then save it as a template for future use.

■ There are two methods to create a sheet set:
- **Using an example sheet set**: An example sheet set is a premade sheet set that provides the organizational structure of the sheet set, which depends on the discipline. You create sheets, add views, and scale them using an example sheet set.
- **Using existing drawings**: Existing drawings can also be used. The layouts in the drawings are used as sheets, and the folders as subsets.

- There are pros and cons for each method. The first method is a lengthy process, but will give you full control over all the elements. The second method is also tedious since you have to create the layouts twice, once for the drawings and then for the sheet sets, but can also provide a certain amount of control.

23.2 WORKING WITH THE SHEET SET MANAGER PALETTE

- In this section, we will discuss:
  - How to open and close an existing sheet set
  - How to work with the Sheet Set Manager palette

23.2.1 How to Open and Close an Existing Sheet Set

- To show the Sheet Set Manager palette go to the **View** tab, locate the **Palettes** panel, and then select the **Sheet Set Manager** button:

- Or, if there are no opened files, you can click the same icon in the Quick Access menu. Either way, you will see the following palette:

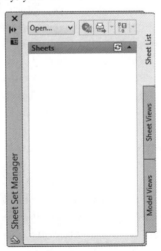

- From the list at the top select Open:

- An Open dialog box will appear. Select your desired *.*DST* file to open it. There are several sample DST files in the following path: \Autodesk\ AutoCAD 2014\Sample\Sheet Sets\. You will find three folders: Architectural, Civil, and Manufacturing.
- Something similar to the following will appear:

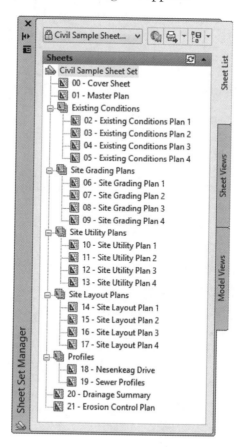

■ To close an existing sheet set click the name of the sheet set in the Sheet Set Manager, then right-click, a menu will appear, and you will select the first option, **Close Sheet Set**:

23.2.2 How to Work with the Sheet Set Manager Palette

■ As shown above, the Sheet Set Manager palette has three tabs:
  • Sheet List tab
  • Sheet Views tab
  • Model View tab
■ While you are in the Sheet List tab (default tab), you will see the following:

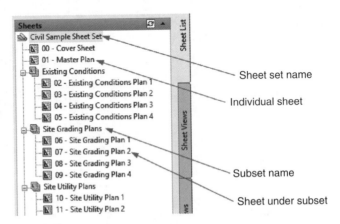

- In the Sheet List tab, you will see the following:
  - Sheet Set name
  - Individual sheets (not listed beneath Subsets)
  - Subsets (method to group sheets)
  - Sheets under Subsets
  - You will see the sheets sorted from top to bottom.
  - For each sheet, there is a number and a title (for instance, 01 is the number for Master Plan sheet).
- We will cover the Model View tab first, then tackle the Sheet Views tab. When you click it you will see something like the following:

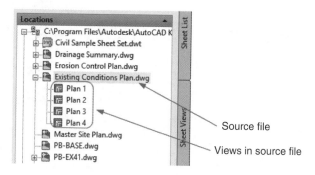

- The Model View tab contains the AutoCAD files called source files and contains the views that will be used in the first method of creating sheet sets (you will use the drag-and-drop method to insert these views as an external reference to the current file).
- If you click on Sheet Views, something similar to the following will be shown:

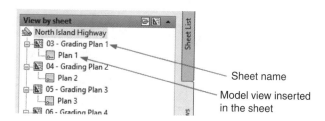

- Using the Model Views tab, if you drag-and-drop a model view, the Sheet View tab will show the name of the sheet, and the name of the view inserted in it.

- ■ While you are in any of the three tabs, you will always be able to see a preview of the file, sheet, and view. See the following two examples.
  - • In the Sheet List tab, if you point (without clicking) on any of the sheets, you will see something like the following:

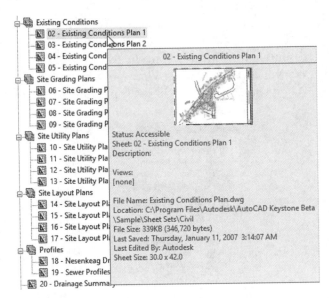

  - • While you are at the Model View tab, if you point to one of the views, you will see something like the following:

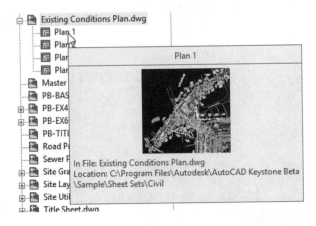

23.3 SHEET SET FILE SETUP

- As discussed previously, AutoCAD will create a DST file that will hold the association and information that defines the sheet set. The DST file does not contain the files and views, but rather the subsets and links to views (or layouts) of the source files. When you create the sheet set AutoCAD will ask you about the location of the DST file along with template files. When the sheet set is created, AutoCAD will create for each sheet a DWG file containing the sheet and the model views inserted in it. If you go to the folder that contains your DST file, you will see something like the following:

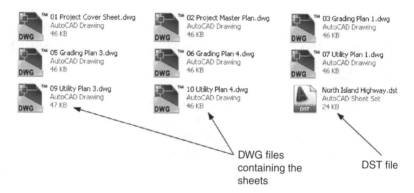

DWG files containing the sheets

DST file

PRACTICE 23-1

Opening, Manipulating, and Closing a Sheet Set

1. Start AutoCAD 2014.
2. Open the Sheet Set Manager palette (regardless of whether a file is open or not).
3. Use the Open command to open the following DST file (depending on the installation of AutoCAD on your machine): \Autodesk\AutoCAD 2014\Sample\Sheet Sets\ Civil\Civil Sample Sheet Set.dst.
4. Double-click sheet 00-Cover Sheet to open it.
5. Close it without saving.

6. Point to sheet 06, which will show more information about it. What is the sheet size? _____

7. Go to Model Views tab.

8. Click the plus sign to expand the list.

9. Click the Existing Conditions Plan.dwg. How many views are there? _____

10. Point to several files (or views) to see a preview of each of them.

11. Double-click Erosion Control Plan.dwg. Did you open Model space or layout? _____

12. Close the file without saving.

13. Go to the Sheet List tab, and close the Civil Sample Sheet Set.

23.4 CREATING A SHEET SET USING THE EXAMPLE SHEET SET

- This is the first of two methods used to create a new sheet set. The merit of this method is that you have full control over it, but this method is lengthy and it takes a long time to develop a sheet set this way.
- In order to create a new sheet set, make sure that at least one file (even if it is empty) is open.
- Open the Sheet Set Manager palette, and from the menu at the top select the **New Sheet Set** option, and you will see the following dialog box:

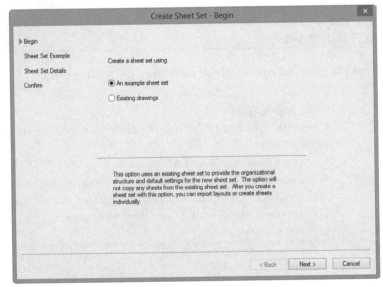

- The default option will be an example sheet set. Click Next to go to the second step, and you will see the following dialog box:

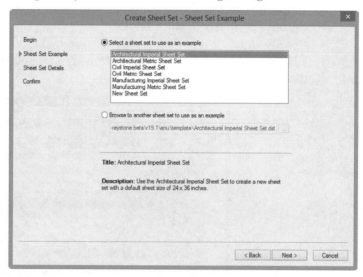

- There are six examples, two for architectural (metric and imperial), two for civil (metric and imperial), and two for manufacturing (metric and imperial). These examples contain only subsets using templates that come with AutoCAD. Another option is to use a previously user-created sheet set (which can be the company's template). When done, click Next to go to the third step, and you will see the following dialog box:

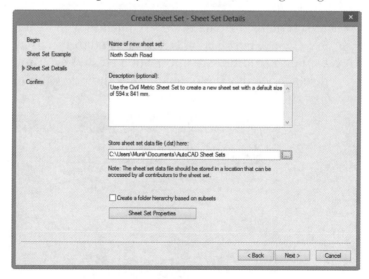

- You should do the following:
  - Input the sheet set title
  - Input the sheet set description
  - Specify the folder to store the DST file
  - Specify whether to create a folder hierarchy based on subsets
  - Click the Sheet Set Properties button, and you will see the following dialog box:

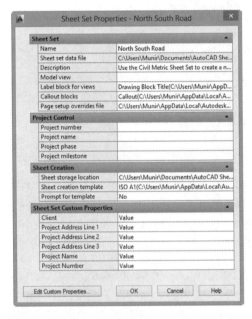

- Almost all of the data is already filled in with default values. The Project Control information is fields related to sheet set, and the Sheet Set Custom Properties are information created by AutoCAD. You can add, delete, or modify these values. Click the Edit Custom Properties button to see the following dialog box:

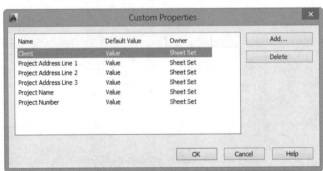

- Here, there are two different pieces of information you should consider changing:
  - Sheet storage location: this piece of information allows you to specify the location of the sheets generated by you. It is best to make it the same folder as the DST file.
  - Sheet creation template: this piece of information allows you to specify the location of the DWT file, and the sheets will be created using it.
- When done, click OK to end Sheet Set Properties, then click Next to go to the fourth step.
- You will see the following dialog box:

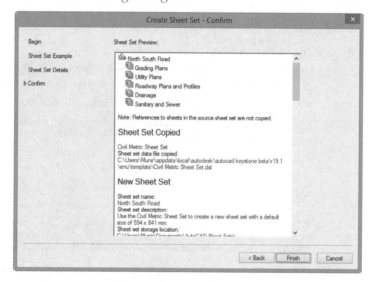

- This is the confirmation page where you will review your settings. Click Finish when you are done. You will be back at the Sheet Set Manager palette and see something similar to the following:

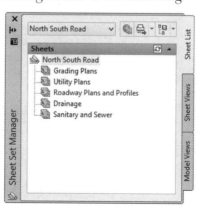

23.4.1 Adding a Sheet in a Subset

- To add a sheet in a subset, simply right-click the name of the subset, and you will see the following menu:

- In this menu, you can:
  - Create a New Sheet in the current subset. You will see the following dialog box; fill in the sheet number and title, then click OK. This will create a new sheet using the template file designated in the creation process:

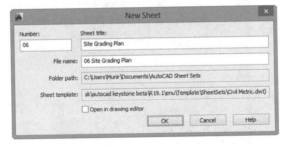

  - Create a New Subset inside the current subset. You will see the following dialog box; fill in the subset name, sheet location, and template:

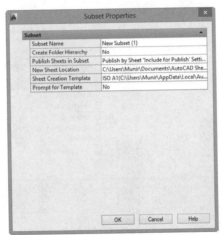

- Import Layout as Sheet
- Rename Subset
- Remove Subset
- Publish
- Properties (which is the same as you used to create it)

23.4.2 Sheet Control

- If you right-click a sheet, you will see the following menu:

- Using this menu you can:
  - Open the sheet (double-clicking the sheet will do the same thing)
  - Open the sheet as read-only
  - Create New Sheet (in the same subset)
  - Rename and Renumber a sheet
  - Remove Sheet
  - Publish the current sheet
  - Use eTransmit
  - View the Properties of the current sheet

PRACTICE 23-2

Creating a Sheet Set Using an Example Sheet Set

1. Start AutoCAD 2014.
2. Make sure a new file is open.
3. Start Sheet Set Manager, and create a new sheet set using an example sheet set.

4. Select the Civil Imperial Sheet Set.
5. Name the new sheet set "North Island Highway."
6. Save the DST file in Practices\Chapter 5\Sheet Sets\Practice 23-1,23-2.
7. Click Sheet Set Properties, and input the following:
   a. Project Number = HW-03
   b. Project Name = Designing North Island Highway
   c. Project Phase = 03
   d. Client = North Island Governorate
8. Finish the creation process.
9. Click the name of the sheet set, then right-click and add the following two sheets:
   a. Sheet Number 01 – Title = Project Cover Sheet
   b. Sheet Number 02 – Title = Project Master Plan
10. Move the two new sheets to the top of the list (using drag-and-drop).
11. Remove the following subsets: Drainage and Sanitary and Sewer.
12. Under the Grading Plans subset, create the following sheets:
    a. 03 – Grading Plan 1
    b. 04 – Grading Plan 2
    c. 05 – Grading Plan 3
    d. 06 – Grading Plan 4
13. Under the Utility Plans subset, create the following sheets:
    a. 07 – Utility Plan 1
    b. 08 – Utility Plan 2
    c. 09 – Utility Plan 3
    d. 10 – Utility Plan 4
14. Using My Computer, go to the folder where you saved the DST file and look at the files created by the sheet set. How does their naming and numbering compare to the sheets? _____
    _____
    _____
15. Leave the Sheet Set Manager open.

23.5 ADDING AND SCALING MODEL VIEWS

- The next step is to fill the sheets with model views from source files. Model views should be prepared prior to this step. Our mission will entail four steps:
  - Specify folder(s) that contain the files holding the desired model views
  - Open the desired sheet

- Drag-and-drop the views in the sheets
- Scale views
■ Model views are considered external references in the sheet, so any change in the original file will be reflected in model views as well.
■ To accomplish the first step, select the Model Views tab, and you will see something like the following:

■ Double-click the Add New Location button to open a dialog box to specify the desired folder. You can repeat this step as many times as needed to add more folders. Once you are done, you will get something like the following:

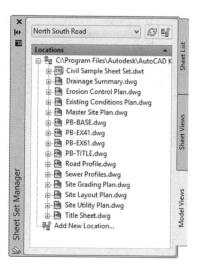

- At the left of some of the files, there is a plus sign. Click it to expand it to see the related model views, just like the following:

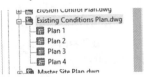

- Open the desired sheet by double-clicking it, and then drag the desired view inside it. Before you decide the exact insertion point, right-click, and you will see a list of scales to be used:

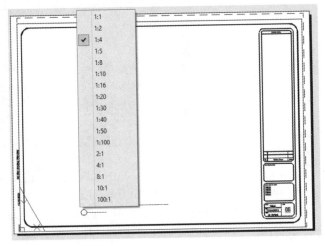

- Select the desired scale, and then specify the insertion point. You should get something like the following:

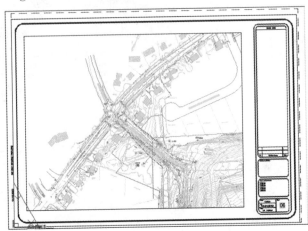

- The view you add in the sheet will be considered the sheet view. To see this, select the Sheet Views tab to see something like the following:

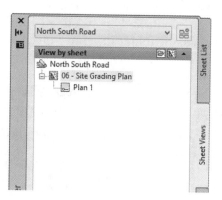

PRACTICE 23-3

Adding and Scaling Model Views

1. Start AutoCAD 2014.
2. Make sure that Sheet Set Manager is open, and the North Island Highway sheet set is open.
3. Double-click sheet number 03, named Grading Plan 1, and notice how a lock is displayed beside it to indicate that this sheet is in use.
4. Go to Model Views tab, double-click Add New Location, and go to \Program Files\Autodesk\AutoCAD 2014\Samples\Sheet Sets\Civil.
5. A set of files will be listed. Click the plus sign beside the Site Grading Plan.dwg file, and you will see four model views. Drag the Plan view into the current sheet and scale it 3/16"=1'.
6. Close and save the file.
7. Do the same thing for the following:
   a. Plan 2 view inside sheet # 04 (same scale)
   b. Plan 3 view inside sheet # 05 (same scale)
   c. Plan 4 view inside sheet # 06 (same scale)
   d. Plan 1 view in Site Utility Plan.dwg inside sheet # 07 (same scale)
   e. Plan 2 view in Site Utility Plan.dwg inside sheet # 08 (same scale)
   f. Plan 3 view in Site Utility Plan.dwg inside sheet # 09 (same scale)
   g. Plan 4 view in Site Utility Plan.dwg inside sheet # 10 (same scale)
8. Close the sheet set.

23.6 CREATING A SHEET SET USING EXISTING DRAWINGS

- If you don't want to create your layouts twice, you can let AutoCAD take all of your layouts and create sheets from them. You can also ask Auto-CAD to take the folder structure you created to be your subset structure. With this method, whatever is in the layout will be considered part of the sheet, but the viewport will not be the external reference.
- While the Sheet Set Manager palette is open choose the **New Sheet Set** option, and the following dialog box will appear:

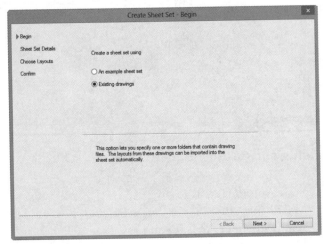

- Select the **Existing drawings** option, click **Next**, and you will see the following dialog box:

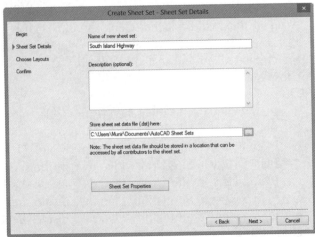

- You should do the following:
  - Input the sheet set title
  - Input the sheet set description
  - Specify the folder to store the DST file in
  - Click the Sheet Set Properties button, and you will see the following dialog box:

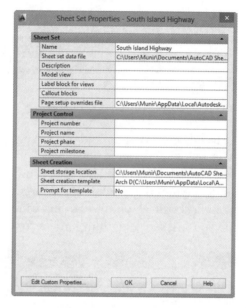

- These are the same properties discussed in the first method. Click OK, and then click Next, and you will see the following dialog box:

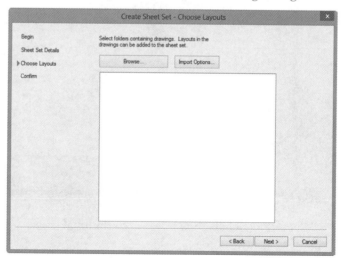

- To control the importing of layouts, click the Import Options button to see the following dialog box:

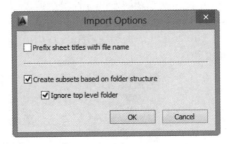

- Control all or any of the following:
  - Prefix sheet titles with file name?
  - Create subsets based on folder structure?
  - Ignore top level folder?
- When done click OK, then click the Browse button to select the folder containing the files that contain the desired layouts. You will then be able to pick and choose which files/layouts will be included in the importing process. You will see something similar to the following:

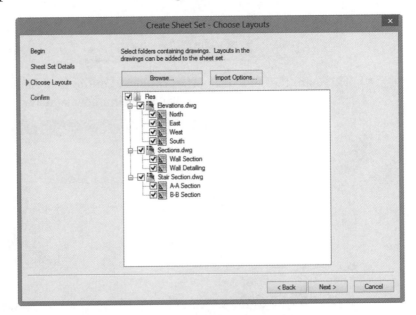

- When done, click Next to jump to the final step.

- You will see the Confirm page. Click Finish to finish the creation process:

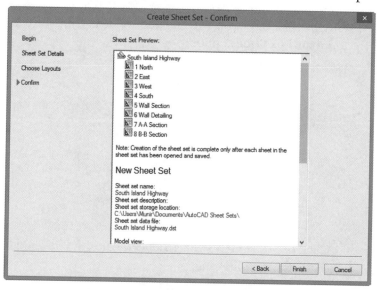

- The Sheet Set Manager will look something like the following:

- As you can see we brought in only sheets, so we need to create subsets to arrange them using drag-and-drop. Check out the following illustration:

- Another helpful tool that can be used with both the first and second method is importing layouts as sheets. Go to the desired subset, then right-click, and you will see the following menu. Choose the **Import Layout as Sheet** option:

- Click Browse for Drawings, select the desired file, and you will see something like the following:

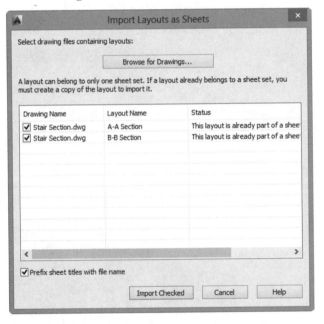

- In the example above, you selected two files, and each contains one layout; you can import either layout, or both of them. Click the Import Checked button to end the importing process.

PRACTICE 23-4

Creating a Sheet Set Using Existing Drawings

1. Start AutoCAD 2014.
2. Make sure that there is at least one empty file open.
3. Create a new sheet set titled "Mira House" using an existing drawing. Save it in Practice\Chapter 5\Sheet Sets\Practice 23-4.
4. You will find the drawings that contain the desired layouts in Practice\ Chapter 5\Sheet Sets\Practice 23-4\Res.
5. Uncheck the Stair Section.dwg file, taking with you only layouts from Elevations.dwg and Sections.dwg.
6. Create three subsets: Evaluations, Sections, and Stair Sections.
7. Using drag-and-drop move North, East, West, and South to the Elevations subset, and Wall Section and Wall Detailing to Sections.
8. Right-click the Stair Sections subset and select Import Layout as Sheet.
9. Browse to the folder Practice\Chapter 5\Sheet Sets\Practice 23-4\Res and select Stair Section.dwg.
10. Import both layouts.
11. Renumber and rename the two imported sheets to be 7 – A-A Section and 8-B-B Section.
12. Close the sheet set.

23.7 PUBLISHING SHEET SETS

- Using Sheet Set Manager, if you right-click the name of the sheet set, a menu will appear with the **Publish** option. You will see the following submenu:

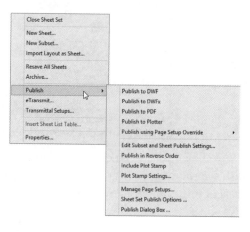

- The first four options were discussed previously: Publish to DWF, Publish to DWFx, Publish to PDF, and Publish to Plotter. The fifth, Publish using Page Setup Override, is new, and it will help you override the default page setup. In order for this option to be available, you should set the Page setup overrides file in the Properties dialog box, as in the following illustration:

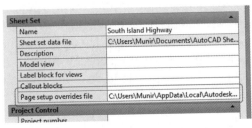

- If this option is properly set, you will see the following:

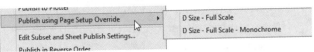

- Below is the Edit Subset and Sheet Publish Settings, which will show the following dialog box:

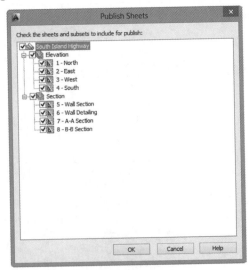

- This box allows you to specify the subset/sheet to be included using the Publish command.
- The rest of the options have already been discussed, or are self-explanatory.

23.8 USING THE eTRANSMIT AND ARCHIVE COMMANDS

- AutoCAD provides two methods to exchange and archive sheet sets, the eTransmit and Archive commands. Both commands will group all the related files and package them in a single file.

23.8.1 Using the eTransmit Command

- To create a package of all files of the sheet set, simply right-click the name of the sheet set, and you will see the following menu:

- You will see the following dialog box:

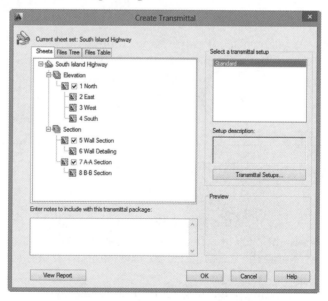

- This dialog box is identical to the one discussed in Chapter 22, but with only one change: the Sheet tab at the upper left of the dialog box.

23.8.2 Using the Archive Command

- To issue this command, right-click the name of the sheet set, and a menu will appear; select the Archive command. You will see the following dialog box:

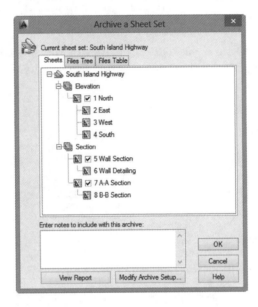

- As you can see it is the same as the eTransmit dialog box. The Archive command will not save a setup like we do using the eTransmit command, rather, you have to use Modify Archive Setup each time you issue it.

PRACTICE 23-5

Publishing and eTransmitting a Sheet Set

1. Start AutoCAD 2014.
2. Open the Mira House sheet set you created in the previous practice.
3. Right-click the sheet set name, select Publish, then Publish to DWF.
4. Name the file Mira House.dwf, and save it in the same folder you saved the DST file in.
5. You can open it using Autodesk Design Review to view it.
6. Right-click the sheet set name again and select eTransmit.

7. Make sure that all sheets are included.
8. Go to the Files Table tab to make sure the Mira House.dst file is included.
9. View the report.
10. Produce and eTransmit the package and save it in the same folder you saved the DST file in.
11. Close the sheet set.

NOTES:

CHAPTER REVIEW

1. How many methods are there to create a sheet set?
   a. One
   b. Three
   c. Two
   d. Four
2. _____ is a method that helps you bring in layouts from other drawings and use them as sheets.
3. Using an example sheet set, you will have viewports ready to use.
   a. True
   b. False
4. Using an example sheet set:
   a. AutoCAD will create sheets
   b. AutoCAD will create subsets
   c. AutoCAD will create both subsets and sheets
   d. AutoCAD will create both subsets, sheets, then add views
5. Using existing drawings, AutoCAD will consider folders as subsets and layouts as sheets.
   a. True
   b. False
6. In the Sheet List tab, if you point (without clicking) on any of the sheets, you will see the _____ of the sheet.

CHAPTER REVIEW ANSWERS

1. c
3. b
5. a

24

CAD STANDARDS AND ADVANCED LAYERS

In This Chapter

◇ What is a CAD standard?
◇ How to create DWS files
◇ What are layer filters?
◇ Using the Layer States Manager
◇ Advanced functions for controlling layers

24.1 WHY DO WE NEED CAD STANDARDS?

- Today's projects are getting larger and more complex, and companies that use AutoCAD (or any other software package) are beginning to realize the need for standards for maintaining continuity across projects and keeping consistent dimensions, text, and layering systems. AutoCAD offers a tool for producing CAD standard files and linking them to DWG files as well as a way to check to make sure files do not violate any of the set standards.
- The process of creating a standard file is lengthy and entails the following steps:
  - Find an international standard or create your own.
  - Sit with all the parties that will be affected by the new standard and get input.
  - Put the standard on paper first and get the approval of all stakeholders.
  - Create a standard file using AutoCAD.
  - Test it and get feedback from users.
  - Take any corrective steps needed.
  - Put the standard in use and monitor it.

- Before this tool was introduced in AutoCAD, CAD managers had to check files manually, one at a time, a tedious process that often introduced errors.
- Another useful tool in AutoCAD is Layer Translator, which will map layers in your current file to compare them to another file to allow you to correct any errors.

24.2 HOW TO CREATE A CAD STANDARDS FILE

- To create a CAD standards file (*.DWS) take the following steps:
  - Open a DWG file containing all the desired layers, dimension styles, text styles, and linetypes. Or, you can create a new file using a template file containing the above.
  - Go to the Application menu, select Save As, and then the **AutoCAD Drawing Standards** option:

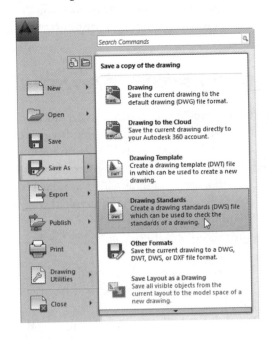

- You will see the following dialog box:

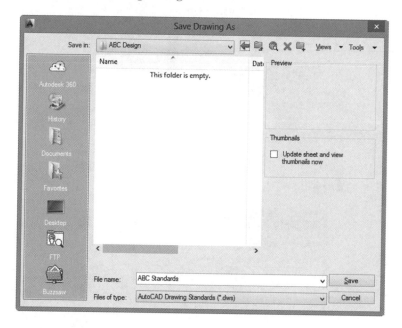

- As you can see, AutoCAD will use the file extension DWS automatically, so you can input the name of the new standard file and then click the Save button. This process will take place only once.

24.3 HOW TO LINK DWS FILES TO DWG FILES AND CHECK THEM

- After you create the DWS file, you will perform two more steps:
  - Configure (link) it to a DWG file
  - Check it to find any violation of the standards

24.3.1 Configuring (Linking) a DWS File to a DWG File

- To link take the following steps:
  - Open the desired DWG file to be checked.

- Go to the **Manage** tab, locate the **CAD Standards** panel, and then select the **Configure** button:

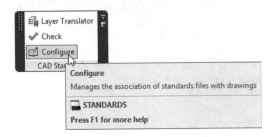

- You will see the following dialog box:

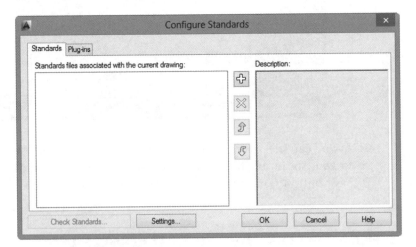

- Click the plus sign to link the current DWG file with the desired DWS file:

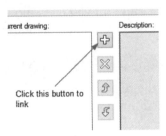

- You will see a file dialog box. Select the DWS file, then click OK.
- To control what to check for, click the Plug-ins tab, and the following will appear:

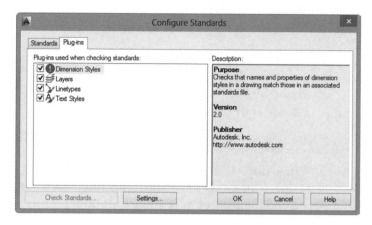

- AutoCAD can only check four things: Dimension, Text Styles, Layers, and Linetypes. To have full control of the process, click the Settings button, and the following dialog box will appear:

- Control the Notification settings by selecting one of the following:
  - Disable standards notifications
  - Display alert upon standards violation (this is used when you are still working with the DWG file)

- Display standards status bar icon (default):

- Control the Check Standards settings and change the following:
  - Automatically use non-standard properties?
  - Show ignored problems?
  - Choose your preferred DWS file

24.3.2 Checking the DWG File

- The first step should be to open the DWG file you want to check for standards compliance. Then go to **Manage**, locate the **CAD Standards** panel, and then select the **Check** button:

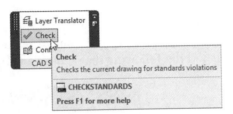

- You will see the following dialog box:

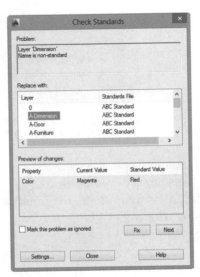

■ As you can see this dialog box is divided into three parts:
  • Problem, where AutoCAD will show the problem that violates the standard. In the example below, the name of a layer in the DWG file is not in the standard file:

  • Replace with, where AutoCAD will propose to replace the mistake with something already in the standard file, as in the following example. The Dimension layer will be replaced with the A-Dimension layer. In this case, AutoCAD will remove the old layer and transfer all the objects in it to the new layer:

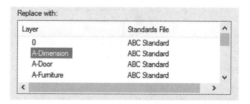

  • Preview of changes, where AutoCAD will show the changes between the old and the new. In the following example, the color will change:

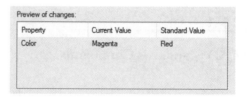

■ AutoCAD offers three solutions to any problem:
  • Fix: To fix the problem using Replace with option.
  • Next: Skip the problem and go to the next problem.
  • Ignore: Ignore this problem and go to the next problem.

- When this command is finished, it will show the following message:

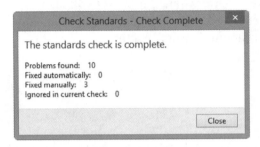

- You can choose to go with another approach, and link the DWS file with an incomplete DWG. This means that when you make any changes to dimensions, text styles, or layers, an instant message at the lower right will appear, something like the following:

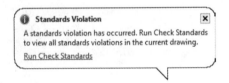

- If you click the Run Check Standards hyperlink at the bottom of the bubble, it will take you directly to the same dialog box discussed above.

PRACTICE 24-1

   Using CAD Standards Commands

1. Start AutoCAD 2014.
2. Open ABC Standard.dwg, and create from it ABC Standard.dws, then close the file.
3. Open **Practice 24-1.dwg**.
4. Using Configure link Practice 24-1.dwg to ABC Standard.dws.
5. Go to the Plug-ins tab, and make sure that all four available options are turned on, and then click OK to end the command.
6. Before you start checking for compliance, check the color of layer Dimension. What is the color? _____

7. Start the Check command, and make the following changes:

Problem	Action
Dim Style: Outside Walls	Replace it with Outside
Dim Style: Inside	Replace it with Inside
Layer: Dimension	Replace it with A-Dimension
Layer: Furniture	Replace it with A-Furniture
Layer: Text	Replace it with A-Text
Layer: Title Block	Replace it with Title Block
Layer: TobeHidden	Replace it with 0
Text Style: Room Titles	Replace it with Room Titles
Text Style: TitleBlock_Bold_Mine	Replace it with TitleBlock_Bold
Text Style: TitleBlock_Regular	Replace it with TitleBlock_Regular
Text Style: Arial_09	Replace it with Standard from DWS file

8. How many problems did AutoCAD find? _____
9. Go to the Layer Properties Manager palette, and create a new layer called "Test." What happened? _____
10. Run Check Standards, and replace Test with 0.
11. Save and close the file.

24.4 USING LAYER TRANSLATOR

- Layer Translator is a tool that will help you translate a file's layering system to your layering system using the following steps:
  - Open the desired DWG file.
  - Link it with a file (DWG, DWS, or DWT) holding your layering system.

- Use Layer Translator to translate layers (naming, colors, linetype, etc.).
- Save the translation in a file, so you can use it for other files.
- To issue this command, go to the **Manage** tab, locate the **CAD Standards** panel, and then select the **Layer Translator** button:

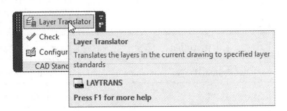

- You will see the following dialog box:

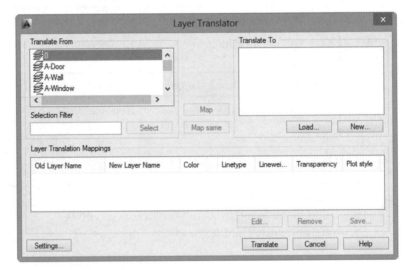

- At the top right, you can see the Translate From list that contains the list of layers that need to be translated to your layering system. The Translate To list is empty because we have not yet added the file. To load your file, click the **Load** button. You can load DWG, DWT, and DWS files.

- This is what you will see after you load your file:

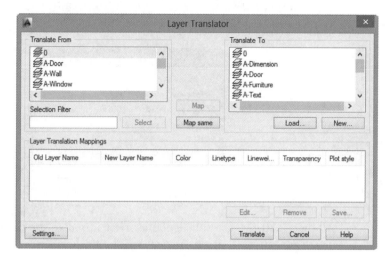

- To find the common layers between the two files click the Map same button in the middle, and you will get something like the following:

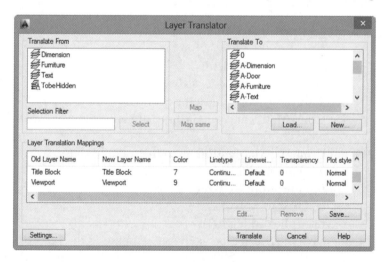

- Whatever is left in the Translate From part are the layers that don't match your file. Select one layer from Translate From, and select one layer from Translate To, then click the Map button. Keep doing that until you finish all the layers. You can also leave some layers unmatched, which means these layers will be left in the file without changing them.

- Use the Save button to save the translation for future use.
- To manipulate the process click the Settings button, and you will see the following dialog box:

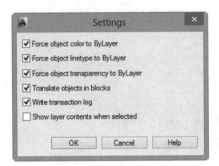

- All these settings are self-explanatory. To finalize the command click the Translate button to perform the translation process.

PRACTICE 24-2

Using Layer Translator

1. Start AutoCAD 2014.
2. Open **Practice 24-2-A.dwg**.
3. Start the Layer Translator command and load ABC Standard.dws.
4. Using Map Same, see if there are any common layers between the two files.
5. Using the Map button, map the following layers:
   a. Dimension to A-Dimension
   b. Furniture to A-Furniture
   c. Text to A-Text
   d. TobeHidden to 0
6. Save the translation as the same name as the file with extension dws.
7. Close Practice 24-2-A.dwg, and open Practice-2-B.dwg.
8. Load the DWS file you saved in step 6.
9. See how all the layers are already translated.
10. Save and close the file.

24.5 WORKING WITH THE LAYER PROPERTIES MANAGER

- AutoCAD gives you all the necessary tools to control the Layer Properties Manager palette. To issue the command, go to the **Home** tab, locate the **Layers** panel, and then select the **Layer Properties** button:

- You will see the following palette:

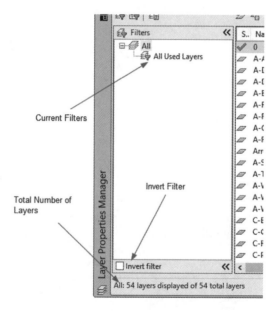

- Note the three arrows that identify the three parts of the palette:
  - Total number of layers will show the used and unused layers in the current drawing.
  - Current filters: By default you will see a single filter, "All Used Layers," which will show only used layers.
  - Invert filter will show the inverse of your current filter.

- While the palette is open, you can do all or any of the following:
  - Hide unneeded columns. Simply right-click one of the columns and a menu will appear. All displayed columns have a (✓) at the left, so to hide any of the columns, click the check:

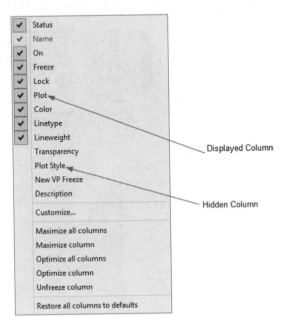

Displayed Column

Hidden Column

- The sort layer list is based on columns and is either ascending or descending. Simply click the title of the column, and an arrow will appear; if it is pointing upwards, the list is ascending, and if it is pointing downwards the list is descending.

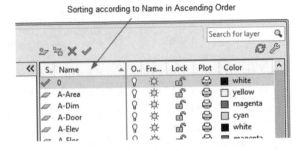

Sorting according to Name in Ascending Order

- Relocate the column position using drag-and-drop:

- Change the column width, or freeze and unfreeze a column
- By right-clicking the heading of a column, the following menu will appear, and at the bottom there are six commands: Maximize all columns (show the heading, or the contents), Maximize column (the one you right-click), Optimize all columns (show the contents, ignoring the heading), Optimize column, and Freeze/Unfreeze column, which means if you horizontally scroll this column it will always be shown. Finally, you can restore all the columns to their defaults:

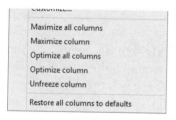

24.6 CREATING A PROPERTY FILTER

- With the complexity of today's drawings, there are often many layers in a single drawing. AutoCAD helps you create a filter to show some of the layers based on a single property or more. Start the Layer Properties Manager, then click the New Property Filter button:

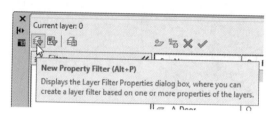

- You will see the following dialog box. Take the following steps:
  - Type the name of the filter.
  - Under Filter definition, use the fields to input one criteria or more. In the following example, we used two criteria, the name (A-*) and the color (color = Cyan):

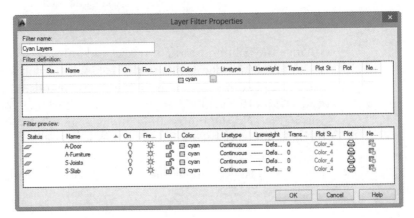

  - AutoCAD will automatically show layer(s) that satisfy the criteria (in our example there are only four layers).
  - When you click OK and go back to the Layer Properties Manager you will see that the new property filter has been added, just like the following:

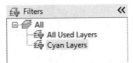

- At the lower left corner, you will see the following:

- At the same time, the layer list in the Layers panel will show only the layers of the filter:

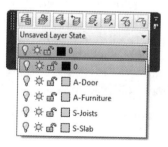

24.7 CREATING A GROUP FILTER

- A group filter is a filter used to group layers that have nothing in common.
- While the Layer Properties Manager is open click the New Group Filter button:

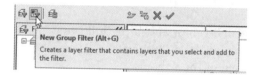

- A new filter will be added; type your desired name. Initially this filter will be empty. Click All filter, then using drag-and-drop add the desired layers:

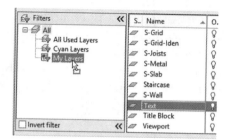

24.8 THINGS YOU CAN DO WITH FILTERS

- One advantage of filters is that you can show some of the layers and hide others, and AutoCAD allows you to perform certain actions on all the layers in the filter in one step.
- When you right-click the filter, a special menu will appear for each type.

24.8.1 Property Filter Menu

- Right-click the property filter, and you will see the following menu:

- You can do all or any of the following:
  - Change the visibility of the layers. You can turn them on or off, freeze them or thaw them.
  - Lock or unlock the layers.
  - If you are in a layout, you can choose to freeze or thaw layers in the current viewport.
  - If you are in a layout, you can choose to isolate layers in all the viewports, or active viewport only.
  - Create a new property or group filter.
  - Convert the property filter to a group filter.
  - Rename and delete a property filter.
  - Change the criteria the filter uses by selecting Properties. This option will show you the same dialog box you use to create the property filter.

24.8.2 Group Filter Menu

- Right-click the group filter, and you will see the following menu:

- Almost all of the options were discussed in the section on the property filter menu, except instead of having the Properties option we have the Select Layers option. You will see the following submenu:

- This option allows you to add/replace more layers to a group filter using the normal selecting method. You will see the following prompt:

```
Add layers of selected objects to filter...:
```

- Click the desired object, and the layer containing this object will be added to the group filter.

PRACTICE 24-3

Advanced Layer Features and Filters

1. Start AutoCAD 2014.
2. Open **Practice 24-3.dwg**.
3. Open the Layer Properties Manager.
4. Layers are sorted by name (ascending). Sort them by color descending. What is the name of the first layer in the list? _____

5. Maximize all column widths.
6. Show the Description column and the hide Plot Style column.
7. Put the Freeze column to the left of the On column.
8. Unfreeze the Name column.
9. Create a new property filter and name it "Architectural Layers with Cyan color" and set the proper criteria for this filter.
10. Create a new group filter, and add Text, Title Block, Frame, and Viewport.
11. Right-click the group filter and choose the Select Layers/Add option, and select the arrow on the stairs. Check the filter. Is there a new layer? What is the name? _____
12. Right-click the property filter and freeze the layers in it.
13. Save and close the file.

24.9 CREATING LAYER STATES

- A layer state is used to save and retrieve a set of layers with their current state from color, linetype, lineweight, on/off, freeze/thaw, etc. If you saved a certain state, you can retrieve all the states of the layer in one step. There are several ways to use this feature:
  - Make sure the Layer States Manager is displayed, then click the **Layer States Manager** button:

  - Or, using the Layers panel, use the first list, and click the **New Layer State** button:

- Either way, the following dialog box will appear:

- To create a new layer state, click the New button, and you will see the following dialog box:

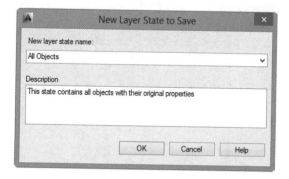

- You should input the name of the new layer state and a good description for its contents. When you click OK, the current state of layers will be saved. To see what properties are saved click the arrow at the lower right of the dialog box, and you will see something like the following:

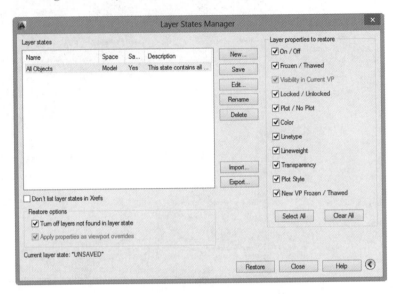

- As you can see, AutoCAD will save all the layer properties. If you click Edit, you will have the choice to alter the state of the layer saved in the layer state, and you will see something like the following:

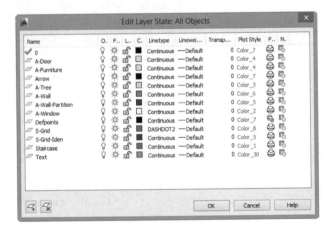

- The rest is self-explanatory.
- To retrieve layer states:
  - You can use the Restore button in the Layer States Manager.

- You can use the first list in the Layers panel:

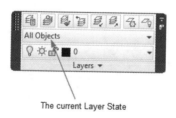

The current Layer State

PRACTICE 24-4

Using Layer States

1. Start AutoCAD 2014.
2. Open **Practice 24-4.dwg**.
3. Start the Layer States Manager, and create a new state from the current state and call it "All Objects," typing the following as the description: "This state contains all layers with their original properties."
4. Freeze layer A-Tree, A-Furniture, and change the A-Window layer's color to blue.
5. Save this state under the name "No Trees and Furniture," and the description as follows: "All Objects except Trees and Furniture."
6. Restore the All Objects state.
7. Using the Edit button, edit No Tree and Furniture by freezing layer Arrow and changing the color of A-Walls-Partition to Cyan.
8. Switch between the two states.
9. If you have time, create your own state.
10. Save and close the file.

24.10 THE SETTINGS DIALOG BOX

- The Settings dialog box allows you to control three things related to layers: what to do when a new layer is added, the settings for isolated layers, and some dialog settings. To display this dialog box, make sure that the Layer Properties Manager is displayed, then click the **Settings** button:

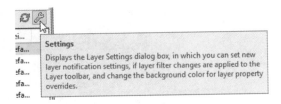

- You will see the following dialog box:

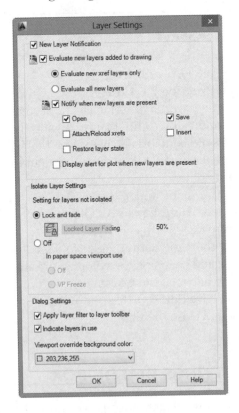

- As you can see there are three parts:
  - New Layer Notification
  - Isolate Layer Settings
  - Dialog Settings

24.10.1 New Layer Notification

- In this part, AutoCAD wants to know whether or not to inform you when a new layer is added. See the following:

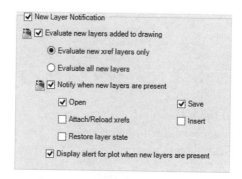

- Control all or any of the following settings:
  - If you want to be notified of new layers
  - Should AutoCAD evaluate all layers, or only xref layers?
  - When the notification should take place: when opening the file, saving the file, attaching/reloading the file, inserting the file in other files, or when you restore a layer state.
  - Should AutoCAD display an alert for plot when new layers are present?

24.10.2 Isolate Layer Settings

- When you issue the Isolate command for layers, how should AutoCAD treat the layers not isolated? There are two choices: either Lock and fade (you should control the fading percentage), or turn isolated layers to be off, and if so, what to do in the layout's viewport (turn them off, or use VP Freeze). See the following:

24.10.3 Dialog Settings

- In this part, you should tell AutoCAD how to deal with several issues related to layers:

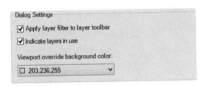

- Control the following:
  - Apply layer filter to layer toolbar?
  - Indicate layers in use:

  - If the layer uses Override setting in viewport, how will AutoCAD distinguish it from other layers? By giving it a background color in the Layer Properties Manager? If so, set the color.
- Now what will happen if a new layer is introduced in a drawing? Do you want AutoCAD to notify you? AutoCAD will show you a bubble just like the following, telling you that a new layer was found, and you need to reconcile it:

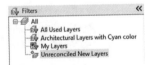

- When you go to the Layer Properties Manager, in the Filters part you will see the following:

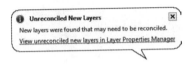

- There is a new filter called Unreconciled New Layers, which contains all the new layers. You can right-click it and choose the **Reconcile Layer** option to accept these new layers, or **Delete Layer**, to reject these layers:

PRACTICE 24-5

 Using the Settings Dialog Box

1. Start AutoCAD 2014.
2. Open **Practice 24-5.dwg**.
3. Start the Settings dialog box
4. Turn on the New Layer Notification checkbox, and change the following settings:
   a. Evaluate new layers added to drawing = on
   b. Evaluate all new layers = on
   c. Notify when new layers are present = on
   d. Open & Save = on
   e. Settings for Layer not isolated = Lock and fade, and fading percentage = 70%
5. Start the Isolate command, and select one line from the outer wall, then press [Enter]. What happened to the other layers?_____.
   Get closer to one of the pieces of furniture, and you will notice that all objects are locked.
6. Add a new layer, call it "Test," and save the file. AutoCAD should inform you that a new layer was added and you need to reconcile it.
7. Go to the Layer Properties Manager, and you will find a new filter called Unreconciled New Layers. Right-click it, and choose to reconcile it.
8. Save and close the file.

NOTES:

CHAPTER REVIEW

1. Which one of the following statements is incorrect?
    a. You can use Layer Translator instead of CAD Standards.
    b. A CAD standards file will check Dimension Style.
    c. A CAD standards file will check Text Style.
    d. The file extension for a standards file is DWS.
2. If you right-click the _____ filter, you will have an option to add layers by selecting them.
3. The property filter will group layers that have the same properties.
    a. True
    b. False
4. In Layer Translator, using the _____ button will find common layers between two files.
    a. Map
    b. Map similar
    c. Map same
    d. Map all
5. The Settings dialog box includes a method to notify you if a new layer is already added.
    a. True
    b. False
6. _____ is used to save and retrieve a set of layers with their current state from color, linetype, lineweight, on/off, freeze/thaw, etc.

CHAPTER REVIEW ANSWERS

1. a
3. a
5. a

25 DRAWING REVIEW

In This Chapter

◇ Review of the Publish command
◇ Using Autodesk Design Review
◇ How to read DWF files in AutoCAD and republish them
◇ Checking and comparing DWF files

25.1 INTRODUCTION TO DRAWING REVIEW

- Normally design drawings are checked and modified several times before the final design is reached. Checking may happen in any design stage by any stakeholder including the design engineer, chief engineer, senior manager, owner, draftsman, or engineer, and all involved parties should keep track of all modifications, including who asked for it, and when.
- Previously this was done manually using hard copy printouts of each drawing, and this process was cumbersome, not to mention wasteful.
- AutoCAD now offers a new fully electronic way to do the revision cycle without even leaving your desk and keeps track of all the changes made to a drawing and by whom. This cycle is based on the following items:
  - DWF file produced from AutoCAD.
  - Autodesk Design Review software. This software comes with AutoCAD and can also be downloaded for free from autodesk.com.
  - The Markup Set Manager, a tool in AutoCAD.

- ▪ The process is very simple:
  - Produce the DWF (single sheet or multiple sheets) and send it to the one who will check (checker) it via e-mail.
  - Using Autodesk Design Review the checker will open your DWF and add any comments and remarks. Autodesk Design Review will track all the changes, and the checker will send it back to you via e-mail.
  - Using Markup Set Manager you will be able to open both the DWF and the DWG and make all the changes required. You will be able to re-publish the DWF again with the first revisions, and send it back to the checker via e-mail.
  - The checker will open both files, the original DWF and the first revision, and compare them to make sure that all modifications asked for were done. The checker may ask for more changes, and then send it back to you.
  - And this process continues until the drawing is complete.

25.2 FIRST STEP – CREATING A DWF FILE

- ▪ In Chapter 10, we covered how to do this part. There are three ways to create a DWF file from the current DWG file:
  - Using the **Plot** command and the *DWF6 ePlot.pc3* printer. This method will produce for each layout along with the Model space a separate DWF file.
  - Using the **Batch Plot** (**Publish**) command. This command will produce a single DWF file containing all layouts and Model space for a file or more.
  - Using the **EXPORTDWF** command. This command will produce a single DWF file containing all layouts of the current file.

25.3 SECOND STEP – USING AUTODESK DESIGN REVIEW

- ▪ When you create a DWF using method 2 and 3 above, Autodesk Design Review will open automatically by default. If not, you will find a shortcut on your desktop. Simply double-click it, then use the normal open file procedure, and you will see something like the following:

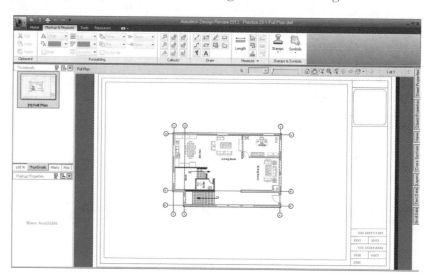

NOTE
- We are using Autodesk Design Review 2013 here, since the newest version wasn't available at the time of production. The next version, however, will likely move to the cloud (i.e. Autodesk 360).
- The Autodesk Design Review interface looks like the AutoCAD interface and uses Ribbons, Application Menu, and the Quick Access Toolbar. At the right, there are seven tabs that will expand when you hover your mouse over them:
  - Sheet Properties, full information about the sheets
  - Views, a list of the named views in the sheets
  - Layers, AutoCAD layers used
  - Text Data
  - Grid Data
  - Cross Sections
  - Object Properties
- Below the ribbon you will see a canvas containing all the viewing commands:

- All these commands are similar to what we learned in AutoCAD. The Home icon will always reset the view to the original view.

- At the left, there are multiple windows:
  - Model will show the Model space if you include it in the DWF file.
  - List Views will show the sheets in a list view.
  - Thumbnails will show the sheets as thumbnails.
  - Markups
  - Markup Properties
- You can combine all windows in one window using drag-and-drop. To do that hold the title of the window and drag it to the list of the tabs at the bottom of the existing window. You can also create a separate window by using drag-and-drop. To do that simply hold the name of the tab at the bottom of the window and drag it outside.

25.4 MARKUP AND MEASURE TAB

- This tab contains all the tools you need to markup and measure the DWF file. It contains six panels:
  - Clipboard
  - Formatting
  - Callouts
  - Draw
  - Measure
  - Stamps and Symbols

25.4.1 Clipboard Panel

- This panel looks like the following:

- Any markup or measure objects you add to the DWF can be selected, so Cut, Copy, and Paste will work.

25.4.2 Formatting Panel

■ Once you start creating markup and measure objects, and just before adding them, you should format them. You will use the following panel:

■ In this panel, there are three parts: the first is used to format the text (one of the text tools should be selected); the second part is used to format lines (one of the other markup and measure tools should be selected to activate this set of tools); and finally, the third part is used to format areas. Here are the text formatting tools:

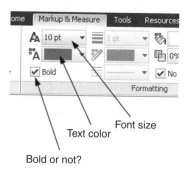

■ Here is the line formatting part:

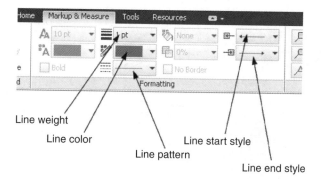

- Lastly, here is the area formatting part:

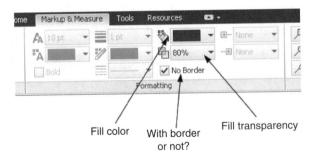

Fill color   With border   Fill transparency
              or not?

25.4.3 Callouts Panel

- This panel gives you nine different shapes for callouts. Six of the nine have a cloud to help you highlight the desired area. It is best to format text, line, and area before inserting the callout. You will see the following panel:

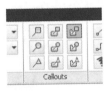

- Here is an example of a callout:

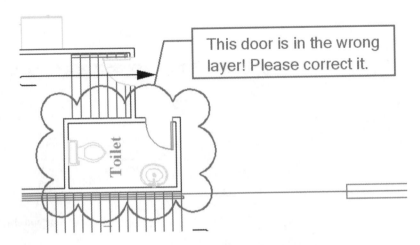

25.4.4 Draw Panel

- The panel looks like the following:

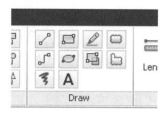

- This panel contains ten different drawing tools:
  - Line
  - Polyline
  - Freehand
  - Rectangle
  - Ellipse
  - Text
  - Freehand Highlighter
  - Rectangle Highlighter
  - Rectangular Cloud
  - Polycloud

25.4.5 Measure Panel

- The panel looks like the following:

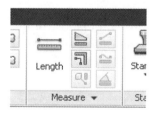

- There are three options to measure 2D DWF files; the rest of the options are for 3D DWF files. These three are:
  - Length, you will need to specify two points in any angle, and you will see a dimension displayed. By default Autodesk Design

Review will snap to an object's end effectively. See the following illustration:

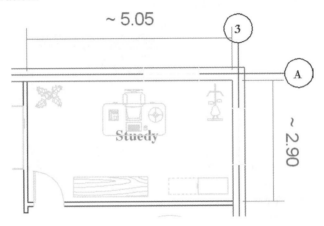

- Area, you will need to specify as many points as desired to get a closed shape to measure its area. See the following illustration:

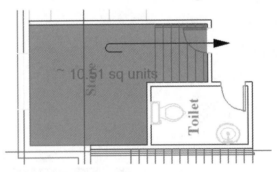

- Polyline, this command will measure the total length of a group of points (you may measure the perimeter of a shape). See the following illustration:

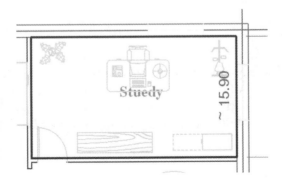

25.4.6 Stamps and Symbols Panel

- This panel looks like the following:

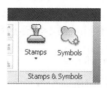

- You can insert seven types of stamps on the DWF file such as Approved, For Review, etc.
- The Symbols button will create a catalog of symbols from the current DWF file or from other DWF files.

25.5 HOW TO EDIT MARKUP OBJECTS

- To edit an existing markup object, take the following steps:
  - From the upper canvas choose the Select tool:

  - Click the objects to be edited.
  - You will see small yellow dots appear at certain places depending on the object selected. Something like the following:

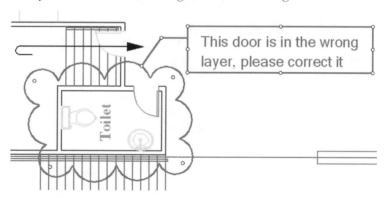

- In this mode, you can simply hit the [Del] key to delete the selected markup object, or move it, resize it (in the above example you can resize the callout alone, or the cloud alone), and you can edit the text.

25.6 CONTROLLING MARKUPS

- As was said at the beginning of this chapter, there will be two windows at the left of Autodesk Design Review: the first one is the Markups window and the second is the Markup Properties window.

25.6.1 Markup Window

- You will see something like the following:

- At the top you will see the name of the sheet containing the markups (in our example it is Full Plan), then you will see a list of the markups added in the current DWF. In the list you will see the type of the markup (rectangle callout, cloud, or text), then the attached text. The list is sorted using the time of inserting. Selecting one of the markups in the list will select the markup object and zoom to it.

25.6.2 Markup Properties Window

- If you select any markup object whether graphically or using the Markups window, go to Markup Properties window, and you will see something like the following:

- In this window, you will see full information on the selected markup such as who created the markup, at what time, and when, as well as the state of this markup, Done, For Review, etc.

PRACTICE 25-1

Creating DWF Files and Using Markup Tools

1. Start AutoCAD 2014.
2. Open **Practice 25-1.dwg**.
3. Create from the Full Plan layout a DWF file and keep the default name.
4. Open the created file using Autodesk Design Review.
5. Measure the area of the store. What is the value? _____ (10.51)
6. Measure the width and length of Study. What are the two values? Width = _____, Length = _____ (5.05, 2.90)
7. From the Callout panel, choose the Rectangle Callout with the Rectangle Cloud tool, and put a callout around the toilet with the following: "This door is not in the right layer."
8. From the Draw panel put a rectangle cloud around the title of room Study, and then add text from the same panel stating the following: "Wrong Spelling and wrong position."
9. From the Callouts panel, select the Rectangle Callout tool, pointing to the upper window of the Study room, and type the following statement: "Window location is wrong."
10. Save the DWF file under the same name and close it.

25.7 USING MARKUP SET MANAGER

- After review, the DWF file should go back to the creator of the DWG file to make the necessary corrections based on the DWF file. In order to do that, you should take the following steps after receiving the DWF file:
  - Start AutoCAD.
  - Open your original DWG file, and go to Model space.

- Go to the **View** tab, locate the **Palettes** panel, and then select the **Markup Set Manager** button:

- When the Markup Set Manager opens, you will see the following palette:

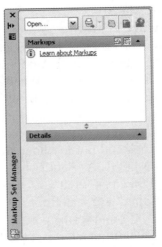

- From the upper pop-up list, select the **Open** option, select your desired DWF file, and the palette will change to something like the following:

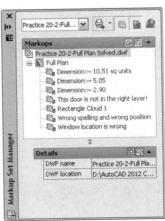

- You will see the name of the DWF file, the name of the layout, and all of the markups listed. Selecting one the markups will show details below the list of the markups, just like the following:

- As you can see, you will see Markup status, Markup creator, Markup created date and time, DWF status, and Markup history, and also Notes.
- In order to see the markups on the DWG drawing, simply double-click the markup. This will take you to the layout, zoomed to the part of the layout that was marked.
- Go to Model space (or double-click) in the viewport to make the necessary changes.
- After finishing the corrections, go to the Markup Set Manager palette, and change the status of the markup to Done, something like the following:

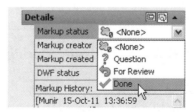

- While both DWG and DWF objects are shown, at the top right of the palette you will see the following:

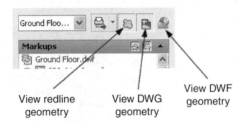

View redline
geometry

View DWG
geometry

View DWF
geometry

- For the two button at the right (i.e., View DWG Geometry and View DWF Geometry), if the first is on, the other is off.
- The third button allows you to turn on/off the markup objects.
- The final step should be to re-publish the DWF file with the new changes. Click the Republish Markup DWF button, and you will see the following selections:

- Select one of the two choices: Republish All Sheets or Republish Markup Sheets only. You can use the Batch Plot command to produce the new revised DWF file.
- When you republish you should give the DWF a new name to indicate the revision has been made, along with a number to make the revision number meaningful. The name should be of this type, *name_rev_01* .dwf, so you can keep track of revisions.

PRACTICE 25-2

Creating DWF Files and Using Markup Tools

1. Start AutoCAD 2014.
2. Open **Practice 25-1.dwg**.

3. Open the Markup Set Manager palette.
4. Open the DWF file you created in the previous exercise.
5. You will see seven markups, but only three need to be fixed; the other four are either measurements or clouds.
6. Double-click the markup the starts with: "This door ...," and this will take you to the Full Plan layout; zoom to the door. Go to Model space, and move the door to the A-Door layer.
7. Do the same for the other two remarks, correcting the misspelled word, moving the text downward little bit, and then moving the window to the left by 1 unit.
8. Change the status of all markup objects to "Done."
9. Using the Batch Plot command, create from the Full Plan layout a DWF and name it Practice 25-1-Full Plan_Revision_001.dwf.
10. Using the Markup Set Manager close the DWF file, then close the palette.
11. Save the DWG file, and close it.

25.8 COMPARING DWF FILES

- Now we will compare the DWF files to make sure all the corrections were made.
- We need to open the old (first) file that contains the markups using Autodesk Design Review, and then open the new one to make the comparison. After opening the old file, go to the **Tools** tab, locate the **Canvas** panel, and then select **Compare Sheets**:

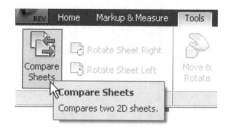

- You will see the following dialog box:

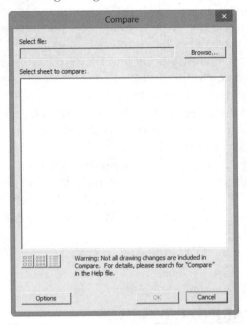

- Click the Browse button, select your new file, and you will get something like the following:

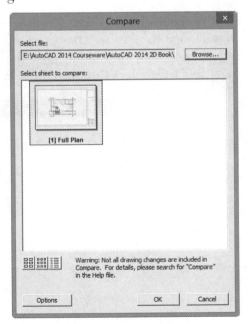

- Click the Options button, and the following dialog box will appear:

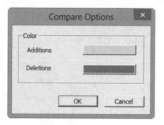

- Additions will be displayed in green and deletions will be displayed in red.
- So you will see the additions and deletions as in the following:

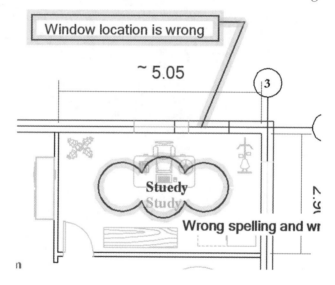

PRACTICE 25-3

Comparing DWF Files

1. Start Autodesk Design Review.
2. Open **Practice 25-1-Full Plan.dwf**.
3. Compare it with Practice 25-1-Full Plan_Revision_001.dwf.
4. Check the three objects that needed to be corrected. Were all corrections done? _____
5. Close Autodesk Design Review.

NOTES:

CHAPTER REVIEW

1. The name of the software that can open and mark DWF files is Autodesk Design Review.
   a. True
   b. False
2. You can _____ two DWF files using Autodesk Design Review.
3. Which one of the following can't be done in Autodesk Design Review?
   a. Calculate an area
   b. Input a dimension of a line using two points
   c. Add some reline objects
   d. Import DWG and DWF files in the same session
4. While you are using Autodesk Design Review you can:
   a. Switch the layers on/off
   b. Check the markup window to see the markup objects included in the current file
   c. Change the layer of a selected object
   d. Check the properties of the markup
5. Using Markup Set Manager you can open a DWF while you are in AutoCAD.
   a. True
   b. False

CHAPTER REVIEW ANSWERS

1. a
3. d
5. a

INDEX